SpringerBriefs in Mathematics

Series Editors

Krishnaswami Alladi
Nicola Bellomo
Michele Benzi
Tatsien Li
Matthias Neufang
Otmar Scherzer
Dierk Schleicher
Benjamin Steinberg
Vladas Sidoravicius
Yuri Tschinkel
Loring W. Tu
G. George Yin
Ping Zhang

SpringerBriefs in Mathematics showcases expositions in all areas of mathematics and applied mathematics. Manuscripts presenting new results or a single new result in a classical field, new field, or an emerging topic, applications, or bridges between new results and already published works, are encouraged. The series is intended for mathematicians and applied mathematicians.

For further volumes:
http://www.springer.com/series/10030

Gabriel N. Gatica

A Simple Introduction to the Mixed Finite Element Method

Theory and Applications

Springer

Gabriel N. Gatica
Centro de Investigación en Ingeniería Matemática
and Departamento de Ingeniería Matemática
Universidad de Concepción
Concepción, Chile

ISSN 2191-8198 ISSN 2191-8201 (electronic)
ISBN 978-3-319-03694-6 ISBN 978-3-319-03695-3 (eBook)
DOI 10.1007/978-3-319-03695-3
Springer Cham Heidelberg New York Dordrecht London

Library of Congress Control Number: 2013958374

Mathematics Subject Classification (2010): 65J05, 65J10, 65N12, 65N15, 65N22, 65N30, 65N50, 74B05, 74F10, 76D07, 76S05

Printed on acid-free paper

Springer is part of Springer Science+Business Media (www.springer.com)

To the memory of Professor Dr. HERMANN ALDER WELLER,
founder of the numerical analysis *discipline at Concepción*
with great affection and enormous gratitude.

Preface

The main purpose of this monograph is to provide a simple and accessible introduction to the mixed finite element method as a fundamental tool to numerically solve a wide class of boundary value problems arising in physics and engineering sciences. The book is based on material that I have used to teach corresponding undergraduate and graduate courses at Universidad de Concepción, Concepción, Chile, during the last 10 years. As compared with several other classic books on the subject, and in addition to being of a limited scope, the main features of the present work concern, on the one hand, my attempt to present and explain most of the details in the proofs and in the various applications. In particular, several results and aspects of the corresponding analysis that are usually available only in papers or proceedings are included here. In addition, keeping in mind that the subject is growing and evolving very quickly, I concentrate the discussion mainly on those core concepts and fundamental results that need to be understood by thesis students and young researchers so that they can read more advanced textbooks and make their own contributions in this and related fields. As a consequence, one of the main emphases of the book is on most of the mathematical and numerical issues involved in the application of the mixed finite element method to simple modeling problems in continuum mechanics. This includes classical Poisson and linear elasticity problems, both under several kinds of boundary conditions for which, among other matters, complete proofs of the continuous and discrete inf-sup conditions required by the theory are provided.

The contents of the book, which assume a basic knowledge of functional analysis, partial differential equations, and Sobolev spaces (e.g., [7, 15, 50, 51, 53, 54]) are described next. Throughout the text, I employ the usual notations from those disciplines, especially the standard terminology for Sobolev spaces. For example, if $\mathcal{O}$ is an open set, its closure, a curve, or a surface, and $s \in \mathbb{R}$, then $\langle \cdot, \cdot \rangle_{s,\mathcal{O}}$, $|\cdot|_{s,\mathcal{O}}$, and $\|\cdot\|_{s,\mathcal{O}}$ denote, respectively, the inner product, seminorm, and norm of the Sobolev space $H^s(\mathcal{O})$. In particular, given Γ, a boundary or part of a boundary, $\langle \cdot, \cdot \rangle_{0,\Gamma}$ represents the inner product of $L^2(\Gamma)$, whereas $\langle \cdot, \cdot \rangle$ stands for the duality pairings of $H^{-s}(\Gamma) \times H^s(\Gamma)$, $H_{00}^{-s}(\Gamma) \times H_{00}^{s}(\Gamma)$, and any vector version of them, for each $s > 0$. However, when it is necessary to identify the underlying Γ, the corresponding duality expression is replaced by $\langle \cdot, \cdot \rangle_\Gamma$. Furthermore, when using the norm $\|\cdot\|_X$ of a given normed space X and when no confusion arises, the subscript $_X$ will usually be omitted. Finally, I use 0 to denote the null scalar as well as the null vector of any space and use C and c, with or without subscripts, bars, tildes, or hats, to denote generic constants independent of eventual discretization parameters, which may take different values at different places.

In Chap. 1, which is of an introductory character, I present a detailed discussion of the classical and general versions of the Lax–Milgram lemma, provide a couple of examples of mixed variational formulations, and prove the main results on traces and Green's identities in $H^1(\Omega)$ and $H(\mathrm{div};\Omega)$. The analysis of the Babuška–Brezzi theory for the aforementioned formulations is the main subject of Chap. 2. The continuous and discrete versions of the theorem, with the necessary and sufficient conditions for unique solvability and the corresponding Cea estimate of the error in the general case, are presented here. In addition, applications to several problems from continuum mechanics, whose respective analyses employ known results from functional analysis and Sobolev spaces, are also provided here. Then, in Chap. 3, I discuss the main facts about the classical Raviart–Thomas spaces. This includes the unisolvency that characterizes their definitions, and the approximation properties of the local and global interpolation operators involved. All the necessary theoretical tools, such as the Denny–Lions and Bramble–Hilbert lemmas and related arguments, are described in this part. Subsequently, after assimilating the contents of this chapter, the reader will easily understand the analysis of similar finite element subspaces available in the literature, such as Brezzi–Douglas–Marini (BDM) and Brezzi–Douglas–Fortin–Marini (BDFM) (e.g., [13, 16]). Finally, specific mixed finite element methods for the boundary value problems discussed in Chap. 2, which consider the Raviart–Thomas finite element subspaces from Chap. 3, are examined in Chap. 4. The corresponding numerical analyses include, among other aspects, the derivation of stable discrete liftings of the associated normal traces, which is particularly relevant for the treatment of Neumann or mixed boundary conditions in three dimensions. The devising of well-posed mixed finite element methods for the linear elasticity problem, which is based on the approach establishing its connection with stable finite element schemes for the usual primal formulation of the Stokes problem, is also discussed briefly in this chapter.

It is time now for the acknowledgements. First of all, I would like to express my deep gratitude to my great collaborators and even greater friends, Salim Meddahi, Norbert Heuer, Francisco J. Sayas, and Antonio Márquez, who, beginning in the late 1990s, and the early, mid, and late following decade, respectively, up to nowadays, have strongly contributed to improving my limited original knowledge of the mixed finite element method and its diverse applications. My deep appreciation also goes to George C. Hsiao for the many fruitful discussions on this and related topics over the years. In addition, I am very thankful to all the undergraduate and graduate students from Universidad de Concepción, Chile, who have taken my regular courses on the subject or have performed their thesis work under my guidance during the last decade. Apologizing in advance for not naming them all, I would like to give special thanks to a former Ph.D. student of mine, Ricardo Oyarzúa, who took the time to read the entire manuscript and pointed out several typographical and mathematical amendments in it. Nevertheless, I am sure that new readers will find more corrections to make, and I thank them in advance for letting me know about the errors. In addition, my gratitude is also due to Mrs. Angelina Fritz, who typeset the original version of the book (written in Spanish) in LaTeX. Finally, I would like to express my appreciation to Springer-Verlag, and especially to Donna Chernyk, Associate Editor of Mathematics, for the publication of this monograph and for the friendly and supportive collaboration along all the way.

This work was partially supported by CONICYT-Chile, through BASAL Project CMM (Universidad de Chile and Universidad de Concepción) and Anillo Project ACT1118 (ANANUM, Universidad de Concepción), and by Centro de Investigación en Ingeniería Matemática (CI^{2}MA), Universidad de Concepción.

Concepción, Chile
October 2013

Gabriel N. Gatica

Contents

Chapter 1
INTRODUCTION

In this chapter we base most of the presentation on the classical references [8, 20, 41, 51] and describe the main introductory aspects of the finite and mixed finite element methods. We first recall the particular and general versions of the Lax–Milgram lemma and then introduce two examples illustrating the use of mixed variational formulations to solve boundary value problems. Finally, we present several basic results on traces, integration by parts formulae, and Green's identities for some Sobolev spaces, and in particular for $H(\mathrm{div};\Omega)$.

1.1 The Lax–Milgram Lemma

To state and prove this result, the most classical one in the analysis of variational problems, we need some preliminary concepts.

1.1.1 Preliminaries

Definition 1.1. Let $(H_1, \langle\cdot,\cdot\rangle_1)$ and $(H_2, \langle\cdot,\cdot\rangle_2)$ be real Hilbert spaces. We say that $B: H_1 \times H_2 \to \mathbb{R}$ is a bilinear form if it is linear in each of its components, that is,

(i) $B(\alpha x + \beta y, z) = \alpha B(x,z) + \beta B(y,z) \quad \forall x,y \in H_1, \quad \forall z \in H_2, \quad \forall \alpha, \beta \in \mathbb{R};$
(ii) $B(x, \alpha y + \beta z) = \alpha B(x,y) + \beta B(x,z) \quad \forall x \in H_1, \quad \forall y,z \in H_2, \quad \forall \alpha, \beta \in \mathbb{R}.$

Definition 1.2. Let $(H_1, \langle\cdot,\cdot\rangle_1)$ and $(H_2, \langle\cdot,\cdot\rangle_2)$ be real Hilbert spaces with induced norms $\|\cdot\|_1$ and $\|\cdot\|_2$, respectively. We say that a bilinear form $B: H_1 \times H_2 \to \mathbb{R}$ is BOUNDED if there exists a constant $M > 0$ such that

$$|B(x,y)| \leq M \, \|x\|_1 \, \|y\|_2 \quad \forall (x,y) \in H_1 \times H_2.$$

G.N. Gatica, *A Simple Introduction to the Mixed Finite Element Method: Theory and Applications*, SpringerBriefs in Mathematics, DOI 10.1007/978-3-319-03695-3_1,

Definition 1.3. Let $(H, \langle\cdot,\cdot\rangle)$ be a real Hilbert space with induced norm $\|\cdot\|$, and let $B : H \times H \to \mathbb{R}$ be a bilinear form. We say that B is STRONGLY COERCIVE (or H-ELLIPTIC) if there exists a constant $\alpha > 0$ such that

$$B(x,x) \ge \alpha \|x\|^2 \qquad \forall x \in H.$$

Now, given $(H_1, \langle\cdot,\cdot\rangle_1)$ and $(H_2, \langle\cdot,\cdot\rangle_2)$ real Hilbert spaces and $B : H_1 \times H_2 \to \mathbb{R}$ a bounded bilinear form, we are interested in defining the operator $\mathbb{B} : H_1 \to H_2$ induced by B and vice versa. To this end, we consider $v \in H_1$ and define the functional $F_v : H_2 \to \mathbb{R}$ by

$$F_v(w) := B(v,w) \quad \forall w \in H_2.$$

Since B is bilinear, it is clear that F_v is linear. In addition, the fact that B is bounded (with constant M) implies that

$$|F_v(w)| \le M\|v\|_1 \, \|w\|_2 \quad \forall w \in H_2,$$

which shows that $F_v \in H_2'$ and

$$\|F_v\| \le M\|v\|_1 \quad \forall v \in H_1. \tag{1.1}$$

The foregoing analysis induces the definition of the operator $\mathscr{B} : H_1 \to H_2'$ as

$$\mathscr{B}(v) := F_v \quad \forall v \in H_1,$$

which, in virtue of the linearity of B in its first component and the inequality (1.1), is linear and bounded with

$$\|\mathscr{B}\|_{\mathscr{L}(H_1,H_2')} \le M.$$

Recall here that, given Banach spaces X and Y, $\mathscr{L}(X,Y)$ denotes the space of bounded linear operators from X to Y. Finally, if $\mathscr{R}_2 : H_2' \to H_2$ denotes the Riesz mapping, we let $\mathbb{B} : H_1 \to H_2$ be the operator induced by B, that is,

$$\mathbb{B} := \mathscr{R}_2 \circ \mathscr{B} \tag{1.2}$$

or, graphically,

$$\begin{array}{ccc} H_1 & \xrightarrow{\mathscr{B}} & H_2' \\ & \searrow_{\mathbb{B}} & \downarrow \mathscr{R}_2 \\ & & H_2 \end{array}.$$

Note that the linearity and boundedness of $\mathscr{R}_2$ and $\mathscr{B}$ yield the same properties for $\mathbb{B}$, and there holds

$$\langle \mathbb{B}(v), w\rangle_2 = \langle \mathscr{R}_2(\mathscr{B}(v)), w\rangle_2 = \mathscr{B}(v)(w) = B(v,w) \quad \forall (v,w) \in H_1 \times H_2. \tag{1.3}$$

Conversely, given $\mathbb{B} \in \mathscr{L}(H_1, H_2)$, we define the bilinear form $B : H_1 \times H_2 \to \mathbb{R}$ induced by $\mathbb{B}$ as

$$B(v,w) := \langle \mathbb{B}(v), w\rangle_2 \quad \forall (v,w) \in H_1 \times H_2. \tag{1.4}$$

1.1.2 Classical Version

The following result constitutes the best known version of the Lax–Milgram lemma.

Theorem 1.1 (Lax–Milgram Lemma). *Let $(H, \langle\cdot,\cdot\rangle)$ be a real Hilbert space, and let $B : H \times H \to \mathbb{R}$ be a bounded and H-elliptic bilinear form with constants M and α, respectively. Then, for each $F \in H'$ there exists a unique $u \in H$ such that*

$$B(u,v) = F(v) \quad \forall v \in H \tag{1.5}$$

and

$$\|u\| \le \frac{1}{\alpha} \|F\|. \tag{1.6}$$

Proof. Let $\mathbb{B} : H \to H$ be the linear and bounded operator induced by B, that is,

$$\langle \mathbb{B}(v), w\rangle = B(v,w) \quad \forall (v,w) \in H \times H,$$

and let $\mathscr{R} : H' \to H$ be the corresponding Riesz operator. Then, finding a unique $u \in H$ such that (1.5) holds is equivalent to looking for $u \in H$ such that

$$\langle \mathbb{B}(u), v\rangle = \langle \mathscr{R}(F), v\rangle \quad \forall v \in H,$$

that is, such that

$$\mathbb{B}(u) = \mathscr{R}(F). \tag{1.7}$$

Moreover, since the foregoing analysis is required for each $F \in H'$, we deduce that the present proof reduces to show that $\mathbb{B} : H \to H$ is bijective. To this end, let us notice from the H-ellipticity of B that for each $v \in H$ there holds

$$\alpha \|v\|^2 \le B(v,v) = \langle \mathbb{B}(v), v\rangle \le \|\mathbb{B}(v)\| \, \|v\|,$$

from where

$$\alpha \|v\| \le \|\mathbb{B}(v)\| \quad \forall v \in H. \tag{1.8}$$

It follows, because of the result characterizing the operators with closed range, that $\mathbb{B}$ is injective and $R(\mathbb{B})$ is a closed subspace of H. Hence, according to the orthogonal decomposition theorem, we have that $H = R(\mathbb{B}) \oplus R(\mathbb{B})^{\perp}$, and therefore, to conclude that $\mathbb{B}$ is surjective, it only remains to prove that $R(\mathbb{B})^{\perp} = \{0\}$. Indeed, given $w \in R(\mathbb{B})^{\perp}$, there holds $\langle z, w\rangle = 0 \quad \forall z \in R(\mathbb{B})$ or, equivalently, $\langle \mathbb{B}(v), w\rangle = 0 \quad \forall v \in H$. In particular, taking $v = w$ and utilizing again the H-ellipticity of B, we obtain

$$0 = \langle \mathbb{B}(w), w\rangle = B(w,w) \ge \alpha \|w\|^2,$$

from which $w = 0$, thus completing the proof of the bijectivity of $\mathbb{B}$. Consequently, given $F \in H'$, there exists a unique $u \in H$ such that $\mathbb{B}(u) = \mathscr{R}(F)$, that is,

$$B(u,v) = F(v) \quad \forall v \in H.$$

Finally, taking $v = u$ in (1.8) and using that $\|\mathbb{B}(u)\| = \|\mathscr{R}(F)\| = \|F\|$, we obtain

$$\|u\| \leq \frac{1}{\alpha}\|F\|,$$

which completes the proof. □

The Lax–Milgram lemma and its proof prompt several remarks. First of all, we observe that the inequality (1.6) represents a continuous dependence result for problem (1.5). In fact, given F_1, $F_2 \in H'$, let us denote by $u_1, u_2 \in H$ the unique solutions, guaranteed by this lemma, of the problems

$$B(u_1, v) = F_1(v) \qquad \forall v \in H$$

and

$$B(u_2, v) = F_2(v) \qquad \forall v \in H.$$

It follows that $u := u_1 - u_2 \in H$ is in turn the unique solution of

$$B(u, v) = (F_1 - F_2)(v) \qquad \forall v \in H,$$

whence (1.6) implies that

$$\|u_1 - u_2\| \leq \frac{1}{\alpha}\|F_1 - F_2\|.$$

The preceding inequality shows that the stability of the solution of (1.5) depends strongly on the inverse of the ellipticity constant α. In other words, the larger α, the better the stability of (1.5).

On the other hand, let us recall from (1.7) that proving the Lax–Milgram lemma reduces, given $F \in H'$, to show the existence of a unique $u \in H$ such that $\mathbb{B}(u) = \mathscr{R}(F)$. Then, considering a parameter $\rho > 0$, the preceding equation is equivalent to finding $u \in H$ such that

$$-\rho\left\{\mathbb{B}(u) - \mathscr{R}(F)\right\} = 0,$$

that is, to seeking $u \in H$ such that

$$T(u) = u,$$

where $T : H \to H$ is the nonlinear operator defined by

$$T(v) := v - \rho\left\{\mathbb{B}(v) - \mathscr{R}(F)\right\} \qquad \forall v \in H.$$

In this way, an alternative proof of the Lax–Milgram lemma consists in proving that T has a unique fixed point, which is achieved, in virtue of the corresponding Banach theorem, by showing that T is a contraction for some $\rho > 0$. In fact, using the ellipticity and boundedness of $\mathbb{B}$, we obtain that

$$\begin{aligned}\|T(v)-T(w)\|^2 &= \langle T(v)-T(w), T(v)-T(w)\rangle \\ &= \langle (v-w)-\rho\,\mathbb{B}(v-w), (v-w)-\rho\,\mathbb{B}(v-w)\rangle \\ &= \|v-w\|^2 - 2\rho\,\langle \mathbb{B}(v-w), v-w\rangle + \rho^2\,\|\mathbb{B}(v-w)\|^2 \\ &\leq (1-2\rho\,\alpha+\rho^2 M^2)\|v-w\|^2 \qquad \forall v,w\in H,\end{aligned}$$

from which it follows that a sufficient condition for the contractivity of T is that $1-2\rho\,\alpha+\rho^2 M^2 < 1$, that is,

$$\rho \in \left(0, \frac{2\alpha}{M^2}\right).$$

Another interesting aspect of problem (1.5) is the resulting analysis for the case where the bilinear form B is SYMMETRIC. Indeed, under this additional hypothesis, B becomes a scalar product on H whose induced norm, denoted by $\|\cdot\|_B$, is given by

$$\|v\|_B := B(v,v)^{1/2} \qquad \forall v \in H.$$

Hence, thanks to the H-ellipticity and boundedness of B, there holds

$$\alpha\,\|v\|^2 \leq B(v,v) = \|v\|_B^2 \leq M\,\|v\|^2 \qquad \forall v\in H,$$

which proves that $\|\cdot\|$ and $\|\cdot\|_B$ are equivalent in H, and therefore $(H, B(\cdot,\cdot))$ is a Hilbert space. Consequently, given $F \in H'$ (with respect to any of these norms), a straightforward application of the Riesz representation theorem (RRT) to $(H, B(\cdot,\cdot))$ yields the existence of a unique $u \in H$ such that

$$F(v) = B(u,v) \qquad \forall v\in H.$$

According to the preceding analysis, the proof of the Lax–Milgram lemma in the case of a symmetric bilinear form B reduces simply to an application of the RRT. In other words, this classical lemma is simply an extension of the RRT to the case of a bounded and H-elliptic bilinear form B.

In what follows, we illustrate the applicability of the Lax–Milgram lemma with a one-dimensional example. To this end and for later use, we recall that, given an interval $\Omega :=]a,b[\subseteq \mathbb{R}$, the corresponding Sobolev space of order 1 is given by

$$H^1(\Omega) := \left\{ v\in L^2(\Omega) : \quad v' \in L^2(\Omega) \right\},$$

where the derivative v' is in the distributional sense. It is easy to prove, using that $L^2(\Omega)$ (with its usual norm $\|\cdot\|_{0,\Omega}$) is Hilbert, that $H^1(\Omega)$ endowed with the inner product $\langle v,w\rangle_{1,\Omega} := \int_\Omega \left\{ v'w' + vw \right\}$ $\forall v, w\in H^1(\Omega)$ and the induced norm $\|\cdot\|_{1,\Omega}$ is also Hilbert. Furthermore, letting $|\cdot|_{1,\Omega}$ be the associated seminorm, that is, $|v|_{1,\Omega} := \|v'\|_{0,\Omega}$ $\forall v\in H^1(\Omega)$, we have the following result.

Lemma 1.1 (Friedrichs–Poincaré Inequality). *Let* $\Omega :=]a,b[\subseteq \mathbb{R}$, *and define* $H_0^1(\Omega) := \left\{ v\in H^1(\Omega) : \quad v(a) = v(b) = 0 \right\}$. *Then there holds*

$$\|v\|_{1,\Omega}^2 \le \left\{1+\frac{(b-a)^2}{2}\right\} |v|_{1,\Omega}^2 \qquad \forall v \in H_0^1(\Omega). \tag{1.9}$$

Proof. We first prove inequality (1.9) in the space $C_0^\infty(\Omega)$ and then use that $C_0^\infty(\Omega)$ is dense in $H_0^1(\Omega)$ with respect to $\|\cdot\|_{1,\Omega}$. In fact, let $\varphi \in C_0^\infty(\Omega)$. Then, for each $x \in \Omega$ there holds

$$\varphi(x) = \int_a^x \varphi'(t)\,dt,$$

which, applying Cauchy–Schwarz's inequality, implies that

$$|\varphi(x)|^2 \le \int_a^x 1^2\,dt \int_a^x (\varphi'(t))^2\,dt = (x-a)\int_a^x (\varphi'(t))^2\,dt$$
$$\le (x-a)\int_a^b (\varphi'(t))^2\,dt = (x-a)\,|\varphi|_{1,\Omega}^2.$$

Then, integrating by parts with respect to $x \in \Omega$ in the preceding estimate, we find that

$$\|\varphi\|_{0,\Omega}^2 \le |\varphi|_{1,\Omega}^2 \int_a^b (x-a)\,dx = \frac{(b-a)^2}{2}\,|\varphi|_{1,\Omega}^2,$$

and therefore

$$\|\varphi\|_{1,\Omega}^2 = \|\varphi\|_{0,\Omega}^2 + |\varphi|_{1,\Omega}^2 \le \left\{1+\frac{(b-a)^2}{2}\right\} |\varphi|_{1,\Omega}^2. \tag{1.10}$$

Now, given $v \in H_0^1(\Omega)$, we let $\{\varphi_n\}_{n\in\mathbb{N}} \subseteq C_0^\infty(\Omega)$ be such that

$$\|v-\varphi_n\|_{1,\Omega} \stackrel{n\to\infty}{\to} 0. \tag{1.11}$$

It follows from (1.10) that

$$\|\varphi_n\|_{1,\Omega}^2 \le \left\{1+\frac{(b-a)^2}{2}\right\} |\varphi_n|_{1,\Omega}^2 \quad \forall n \in \mathbb{N},$$

from which, taking $\lim_{n\to\infty}$ and using (1.11), we conclude (1.9). □

Example 1.1. Given $\Omega =]0,1[$ and $f \in L^2(\Omega)$, we consider the boundary value problem

$$-u'' = f \quad \text{in} \quad \Omega, \quad u(0) = u(1) = 0.$$

It is easy to see that the corresponding variational formulation is given as follows: find $u \in H := H_0^1(\Omega)$ such that

$$B(u,v) := \int_0^1 u'v' = F(v) := \int_0^1 fv \quad \forall v \in H. \tag{1.12}$$

It is clear that F is linear and bounded since, applying the Cauchy–Schwarz inequality, we obtain

$$|F(v)| = \left|\int_0^1 fv\right| \le \|f\|_{0,\Omega}\|v\|_{0,\Omega} \le \|f\|_{0,\Omega}\|v\|_{1,\Omega} \quad \forall v \in H,$$

which says that $\|F\| \leq \|f\|_{0,\Omega}$. Then, $B : H \times H \to \mathbb{R}$ is a bounded bilinear form since, again using the Cauchy–Schwarz inequality, we obtain

$$|B(w,v)| = \left| \int_0^1 w'v' \right| \leq |w|_{1,\Omega}\, |v|_{1,\Omega} \leq \|w\|_{1,\Omega}\, \|v\|_{1,\Omega} \quad \forall\, w,v \in H.$$

On the other hand, employing the Friedrichs–Poincaré inequality from Lemma 1.1 with $a = 0$ and $b = 1$, we find that

$$B(v,v) = \int_0^1 (v')^2 = |v|_{1,\Omega}^2 \geq \frac{2}{3}\, \|v\|_{1,\Omega}^2 \quad \forall\, v \in H,$$

which shows that B is H-elliptic with constant $\alpha = 2/3$. Hence, a direct application of the Lax–Milgram lemma implies that (1.12) has a unique solution $u \in H_0^1(\Omega)$, which satisfies

$$\|u\|_{1,\Omega} \leq \frac{3}{2}\, \|F\| \leq \frac{3}{2}\, \|f\|_{0,\Omega}.$$

1.1.3 General Version

The next goal is to derive a more general version of the Lax–Milgram lemma (Theorem 1.1). To this end, we now consider real Hilbert spaces $(H_1, \langle\cdot,\cdot\rangle_1)$ and $(H_2, \langle\cdot,\cdot\rangle_2)$, a functional $F \in H_2'$, and a bounded bilinear form $B : H_1 \times H_2 \to \mathbb{R}$ and look for $u \in H_1$ such that

$$B(u,v) = F(v) \quad \forall\, v \in H_2. \tag{1.13}$$

Equivalently, if $\mathbb{B} : H_1 \to H_2$ is the linear and bounded operator induced by B, and $\mathscr{R}_2 : H_2' \to H_2$ is the corresponding Riesz mapping, then we are interested in finding $u \in H_1$ such that

$$\mathbb{B}(u) = \mathscr{R}_2(F).$$

Hence, a necessary and sufficient condition for (1.13) to have a unique solution for each $F \in H_2'$ is that $\mathbb{B}$ be bijective. Then, the bijectivity of $\mathbb{B}$ can be reformulated according to the equivalences provided by the following lemma.

Lemma 1.2. *Let $(H_1, \langle\cdot,\cdot\rangle_1)$ and $(H_2, \langle\cdot,\cdot\rangle_2)$ be Hilbert spaces with induced norms $\|\cdot\|_1$ and $\|\cdot\|_2$, respectively, and let $\mathbb{B} \in \mathscr{L}(H_1, H_2)$. Then:*

(a) *$\mathbb{B}$ is surjective if and only if $\mathbb{B}^*$ is injective and has a closed range, that is, if there exists $\alpha > 0$ such that*

$$\|\mathbb{B}^*(v)\|_1 \geq \alpha\, \|v\|_2 \qquad \forall\, v \in H_2. \tag{1.14}$$

(b) *$\mathbb{B}$ is injective if and only if*

$$\sup_{v \in H_2} \langle \mathbb{B}(u), v \rangle_2 > 0 \qquad \forall\, u \in H_1,\ u \neq 0.$$

(c) $\mathbb{B}^*$ *is surjective if and only if* $\mathbb{B}$ *is injective and has a closed range, that is, if there exists* $\alpha > 0$ *such that*

$$\|\mathbb{B}(u)\|_2 \geq \alpha \|u\|_1 \qquad \forall u \in H_1. \tag{1.15}$$

(d) $\mathbb{B}^*$ *is injective if and only if*

$$\sup_{u \in H_1} \langle \mathbb{B}(u), v \rangle_2 > 0 \quad \forall v \in H_2, v \neq 0.$$

(e) $\mathbb{B}$ *is bijective if and only if* $\mathbb{B}^*$ *is bijective.*

Proof.

(a) Suppose that $R(\mathbb{B}) = H_2$. It follows that $R(\mathbb{B})$ and, hence, $R(\mathbb{B}^*)$ are closed. In addition, it is clear that $N(\mathbb{B}^*) = R(\mathbb{B})^\perp = H_2^\perp = \{0\}$. Conversely, if $\mathbb{B}^*$ is injective and has a closed range, the range of $\mathbb{B}$ is closed as well, and therefore $R(\mathbb{B}) = N(\mathbb{B}^*)^\perp = \{0\}^\perp = H_2$. The equivalence with (1.14) is precisely the characterization result for injective operators with a closed range.
(b) It suffices to see that $\mathbb{B}$ is injective if and only if $\mathbb{B}(u) \neq 0 \quad \forall u \in H_1, u \neq 0$.
(c) and (d) These equivalences follow directly from (a) and (b) by applying them to $\mathbb{B}^*$.
(e) Assume that $\mathbb{B}$ is bijective. It follows from (a) that $\mathbb{B}^*$ is injective and has a closed range, which yields $R(\mathbb{B}^*) = N(\mathbb{B})^\perp = \{0\}^\perp = H_1$, and thus $\mathbb{B}^*$ is bijective. For the converse it suffices to apply the preceding implication to $\mathbb{B}^*$ instead of $\mathbb{B}$.

□

It is important to observe here, according to (e), that the pairs of conditions (a), (b) and (c), (d) are equivalent. In this respect, let us also note that (1.14) and (1.15) can be rewritten as

$$\|\mathbb{B}^*(v)\|_1 := \sup_{\substack{u \in H_1 \\ u \neq 0}} \frac{\langle \mathbb{B}(u), v \rangle_2}{\|u\|_1} = \sup_{\substack{u \in H_1 \\ u \neq 0}} \frac{B(u,v)}{\|u\|_1} \geq \alpha \|v\|_2 \quad \forall v \in H_2 \tag{1.16}$$

and

$$\|\mathbb{B}(u)\|_2 := \sup_{\substack{v \in H_2 \\ v \neq 0}} \frac{\langle \mathbb{B}(u), v \rangle_2}{\|v\|_2} = \sup_{\substack{v \in H_2 \\ v \neq 0}} \frac{B(u,v)}{\|v\|_2} \geq \alpha \|u\|_1 \quad \forall u \in H_1 \tag{1.17}$$

or, respectively,

$$\inf_{\substack{v \in H_2 \\ v \neq 0}} \sup_{\substack{u \in H_1 \\ u \neq 0}} \frac{B(u,v)}{\|u\|_1 \|v\|_2} \geq \alpha \tag{1.18}$$

and

$$\inf_{\substack{u \in H_1 \\ u \neq 0}} \sup_{\substack{v \in H_2 \\ v \neq 0}} \frac{B(u,v)}{\|u\|_1 \|v\|_2} \geq \alpha, \tag{1.19}$$

which explains the name of the INF-SUP CONDITIONS given to (1.16) and (1.17) [equivalently, (1.14) and (1.15)].

In virtue of the preceding analysis, we can establish next a more general version of the Lax–Milgram lemma.

Theorem 1.2 (Generalized Lax–Milgram Lemma). *Let* $(H_1, \langle\cdot,\cdot\rangle_1)$ *and* $(H_2, \langle\cdot,\cdot\rangle_2)$ *be Hilbert spaces with induced norms* $\|\cdot\|_1$ *and* $\|\cdot\|_2$, *respectively, and let* $B : H_1 \times H_2 \to \mathbb{R}$ *be a bounded bilinear form. Assume that:*

(i) *There exists* $\alpha > 0$ *such that*

$$\sup_{\substack{v\in H_2\\ v\neq 0}} \frac{B(u,v)}{\|v\|_2} \geq \alpha\,\|u\|_1 \quad \forall\, u \in H_1;$$

(ii)

$$\sup_{u\in H_1} B(u,v) > 0 \quad \forall\, v \in H_2,\, v \neq 0.$$

Then, for each $F \in H_2'$, *there exists a unique* $u \in H_1$ *such that*

$$B(u,v) = F(v) \qquad \forall\, v \in H_2$$

and

$$\|u\|_1 \leq \frac{1}{\alpha}\,\|F\|_{H_2'}. \tag{1.20}$$

Moreover, assumptions (i) *and* (ii) *are also necessary.*

Proof. It suffices to notice that (i) and (ii) correspond to conditions (c) and (d) from Lemma 1.2, which (both together) are equivalent to the bijectivity of $\mathbb{B}^*$ and, consequently, equivalent to the bijectivity of $\mathbb{B}$ as well. Estimate (1.20) follows from (1.15) by noting that $\mathbb{B}(u) = R_2(F)$. □

Certainly, the preceding theorem can be stated, equivalently, with conditions (a) and (b) instead of (c) and (d) from Lemma 1.2. In this case, one can assume the same constant α in (1.14) and (1.15) since $\|(\mathbb{B}^*)^{-1}\|$ [bounded by $1/\alpha$ in (1.14)] is equal to $\|\mathbb{B}^{-1}\|$ [bounded by $1/\alpha$ in (1.15)], and therefore estimate (1.20) is also obtained from (a) and (b), but making use of the identity $\|(\mathbb{B}^*)^{-1}\| = \|\mathbb{B}^{-1}\|$.

Now, in the particular case where $H_1 = H_2 = H$, the Lax–Milgram lemma (cf. Theorem 1.1) follows obviously from Theorem 1.2. Indeed, if $B : H \times H \to \mathbb{R}$ is a bounded and H-elliptic bilinear form with constants M and α, respectively, then there clearly hold

$$\sup_{\substack{v\in H\\ v\neq 0}} \frac{B(u,v)}{\|v\|} \geq \frac{B(u,u)}{\|u\|} \geq \alpha\,\|u\| \qquad \forall\, u \in H,\, u \neq 0,$$

and

$$\sup_{u\in H} B(u,v) \geq B(v,v) \geq \alpha\,\|v\|^2 > 0 \qquad \forall\, v \in H,\, v \neq 0,$$

which show, respectively, hypotheses (i) and (ii) from Theorem 1.2. In turn if instead of being H-elliptic it is assumed that $B : H \times H \to \mathbb{R}$ is symmetric, then the operator $\mathbb{B} \in \mathscr{L}(H)$ induced by B becomes self-adjoint, and hence hypothesis (ii) from Theorem 1.2 is redundant. The preceding analysis suggests the following symmetric version of the generalized Lax–Milgram lemma.

Theorem 1.3. *Let H be a real Hilbert space, and let $B : H \times H \to \mathbb{R}$ be a bounded bilinear form. Assume that:*

(i) $B(w,v) = B(v,w) \qquad \forall\, w,v \in H$;
(ii) *There exists $\alpha > 0$ such that*

$$\sup_{\substack{v\in H\\ v\neq 0}} \frac{B(u,v)}{\|v\|} \geq \alpha\,\|u\| \qquad \forall\, u \in H.$$

Then, for each $F \in H'$ there exists a unique $u \in H$ such that

$$B(u,v) = F(v) \qquad \forall\, v \in H$$

and

$$\|u\| \leq \frac{1}{\alpha}\,\|F\|\,.$$

Proof. It is a straightforward corollary of Theorem 1.2. □

1.2 Examples of Mixed Formulations

1.2.1 A One-Dimensional Model

Let a, b, $\kappa \in \mathbb{R}$, $\kappa > 0$, $\Omega :=]0,1[$, $f \in L^2(\Omega)$, and let us consider the following boundary value problem:

$$-\,u'' + \kappa u = f \quad \text{in} \quad \Omega\,, \quad u'(0) = a\,, \quad u'(1) = b\,. \tag{1.21}$$

The primal formulation of (1.21) is given as follows: find $u \in H := H^1(\Omega)$ such that

$$A(u,v) = F(v) \qquad \forall\, v \in H\,, \tag{1.22}$$

where $A : H \times H \to \mathbb{R}$ is the bilinear form defined by

$$A(u,v) := \int_0^1 \Big\{ u'v' + \kappa u v \Big\} \qquad \forall\, u,v \in H\,,$$

and $F : H \to \mathbb{R}$ is the linear functional given by

$$F(v) := \int_0^1 f v + \Big\{ b v(1) - a v(0) \Big\} \qquad \forall\, v \in H\,.$$

It is important to observe here that the boundary conditions of (1.21) are incorporated automatically, through the integration by parts procedure, into the functional F of the variational formulation (1.22). This is actually a characteristic feature of the Neumann boundary conditions in primal formulations, which explains the name *natural boundary conditions* given to them.

Now, to demonstrate the well-posedness of (1.22), that is, the unique solvability and continuous dependence on the data, it suffices to verify the hypotheses of the classical Lax–Milgram lemma (cf. Theorem 1.1). In fact, since $\kappa > 0$, there holds

$$A(v,v) = \int_0^1 \left\{ (v')^2 + \kappa v^2 \right\} \geq \min\{1,\kappa\} \|v\|_{1,\Omega}^2 \qquad \forall v \in H,$$

which proves that A is H-elliptic. In addition, utilizing the Cauchy–Schwarz inequality, we deduce that

$$|A(u,v)| \leq \max\{1,\kappa\} \|u\|_{1,\Omega} \|v\|_{1,\Omega} \qquad \forall u, v \in H,$$

which shows that A is bounded. For the boundedness of F we observe, also as a consequence of the Cauchy–Schwarz inequality, that

$$\left| \int_0^1 f v \right| \leq \|f\|_{0,\Omega} \|v\|_{0,\Omega} \leq \|f\|_{0,\Omega} \|v\|_{1,\Omega} \qquad \forall v \in H. \tag{1.23}$$

Next, given $v \in C^1(\bar{\Omega})$ (restrictions to Ω of functions that are of class C^1 in an open set containing $\bar{\Omega}$), we have that

$$v(0) = v(x) - \int_0^x v'(t)\,dt \qquad \forall x \in \Omega,$$

whence

$$\begin{aligned} |v(0)|^2 &\leq 2 \left\{ |v(x)|^2 + \left| \int_0^x v'(t)\,dt \right|^2 \right\} \\ &\leq 2 \left\{ |v(x)|^2 + \left(\int_0^x 1\,dt \right) \left(\int_0^x (v'(t))^2\,dt \right) \right\} \\ &\leq 2 \left\{ |v(x)|^2 + x\,|v|_{1,\Omega}^2 \right\} \quad \forall x \in \Omega. \end{aligned}$$

Then, integrating with respect to $x \in \Omega$, we find that

$$|v(0)|^2 \leq 2 \left\{ \|v\|_{0,\Omega}^2 + \frac{1}{2} |v|_{1,\Omega}^2 \right\} \leq 2 \|v\|_{1,\Omega}^2,$$

and hence

$$|v(0)| \leq \sqrt{2} \|v\|_{1,\Omega} \qquad \forall v \in C^1(\bar{\Omega}).$$

Analogously, it is proved that

$$|v(1)| \leq \sqrt{2}\|v\|_{1,\Omega} \qquad \forall v \in C^1(\bar{\Omega}),$$

and finally, the fact that $C^1(\bar{\Omega})$ is dense in $H^1(\Omega)$ allows us to show that both inequalities are extended to H. In this way, it follows that

$$\left| bv(1) - av(0) \right| \leq \sqrt{2}\,(a+b)\,\|v\|_{1,\Omega} \qquad \forall v \in H,$$

which, together with (1.23), shows that F is bounded.

On the other hand, one of the main motivations for using mixed variational formulations, which also constitutes one of the most important features of this methodology, is the possibility of introducing additional variables (unknowns) having either a physical or mathematical interest, which usually depend on the original unknowns.

To illustrate the foregoing principles, let us additionally define $\sigma := u'$ in Ω, so that the boundary value problem (1.21) is reformulated as the first-order system

$$\sigma = u' \ \text{ in } \ \Omega, \quad -\sigma' + \kappa u = f \ \text{ in } \ \Omega, \quad \sigma(0) = a, \quad \sigma(1) = b. \tag{1.24}$$

Note that we now have two unknowns, σ and u, and the Neumann boundary conditions for u become Dirichlet boundary conditions for σ. Then, multiplying the equation $\sigma = u'$ in Ω by $\tau \in H_0^1(\Omega)$ and integrating by parts, we arrive at

$$\int_0^1 \sigma\tau + \int_0^1 u\tau' = 0 \qquad \forall \tau \in H_0^1(\Omega).$$

In addition, multiplying $-\sigma' + \kappa u = f$ in Ω by $v \in L^2(\Omega)$, we obtain

$$\int_0^1 \sigma' v - \kappa \int_0^1 uv = -\int_0^1 fv \qquad \forall v \in L^2(\Omega).$$

In this way, a mixed variational formulation of (1.21) would be given, at first glance, as follows: find $(\sigma, u) \in H^1(\Omega) \times L^2(\Omega)$ such that $\sigma(0) = a$, $\sigma(1) = b$,

$$\begin{aligned} \int_0^1 \sigma\tau + \int_0^1 u\tau' &= 0 && \forall \tau \in H_0^1(\Omega), \\ \int_0^1 \sigma' v - \kappa \int_0^1 uv &= -\int_0^1 fv && \forall v \in L^2(\Omega). \end{aligned} \tag{1.25}$$

However, system (1.25) is not *symmetric* with respect to unknowns and test functions since σ lies in an affine space [translated from $H^1(\Omega)$], and the corresponding test function τ belongs to $H_0^1(\Omega)$. This is caused by the fact that, unlike what happens for a primal formulation, the Neumann boundary conditions are not natural for mixed formulations, which is why in this case they are called *essential boundary conditions*. To circumvent this difficulty, in what follows we proceed in two different ways.

1.2.1.1 Translation of the Unknown σ

Let us define the auxiliary function $\sigma_0(x) := a + (b-a)x \quad \forall x \in \Omega$ and the translated unknown $\tilde{\sigma} = \sigma - \sigma_0$. Note that σ_0 satisfies the boundary conditions from (1.24). It follows that system (1.24) is rewritten as

$$\begin{gathered} \tilde{\sigma} = u' - \sigma_0 \quad \text{in} \quad \Omega\,, \quad -\tilde{\sigma}' + \kappa u = f + (b-a) \quad \text{in} \quad \Omega\,, \\ \tilde{\sigma}(0) = 0\,, \quad \tilde{\sigma}(1) = 0\,. \end{gathered} \tag{1.26}$$

Thus, proceeding as before, the mixed variational formulation of (1.26) reduces to the following: find $(\tilde{\sigma}, u) \in H_0^1(\Omega) \times L^2(\Omega)$ such that

$$\begin{aligned} \int_0^1 \tilde{\sigma}\,\tau + \int_0^1 u\,\tau' &= -\int_0^1 \sigma_0\,\tau & \forall \tau \in H_0^1(\Omega)\,, \\ \int_0^1 \tilde{\sigma}'\,v - \kappa \int_0^1 u v &= -\int_0^1 f v - (b-a)\int_0^1 v & \forall v \in L^2(\Omega)\,, \end{aligned} \tag{1.27}$$

from which it is clear that the unknowns and test functions now reside in the same product space. Unfortunately, this procedure is not applicable, from a practical point of view, to higher-dimensional problems. Indeed, in those cases the existence of a function σ_0 is known, but, in general, it is not possible to obtain it explicitly.

1.2.1.2 Use of a Lagrange Multiplier

Starting from system (1.24), and instead of employing a test function $\tau \in H_0^1(\Omega)$ one simply considers a function $\tau \in H^1(\Omega)$ and introduces the auxiliary unknown (the Lagrange multiplier) $\varphi := (\varphi_1, \varphi_2) \in \mathbb{R}^2$, with $\varphi_1 := u(1)$ and $\varphi_2 := -u(0)$. This induces the weak imposition of the boundary conditions from (1.24) through the simple equation

$$\psi \cdot (\sigma(0), \sigma(1)) = \psi \cdot (a,b) \qquad \forall \psi \in \mathbb{R}^2\,.$$

Consequently, and defining the spaces $H := H^1(\Omega)$ and $Q := L^2(\Omega) \times \mathbb{R}^2$, we arrive at the following mixed variational formulation of (1.24): find $(\sigma, (u, \varphi)) \in H \times Q$ such that

$$\begin{aligned} a(\sigma, \tau) + b(\tau, (u, \varphi)) &= F(\tau) & \forall \tau \in H\,, \\ b(\sigma, (v, \psi)) - c((u, \varphi), (v, \psi)) &= G(v, \psi) & \forall (v, \psi) \in Q\,, \end{aligned} \tag{1.28}$$

where $a : H \times H \to \mathbb{R}$, $b : H \times Q \to \mathbb{R}$, and $c : Q \times Q \to \mathbb{R}$ are the bilinear forms defined by

$$a(\sigma,\tau) := \int_0^1 \sigma\,\tau \qquad \forall\,(\sigma,\tau)\in H\times H\,,$$

$$b(\tau,(v,\psi)) := \int_0^1 \tau' v - \psi\cdot(\tau(0),\tau(1)) \qquad \forall\,(\tau,(v,\psi))\in H\times Q\,,$$

$$c((u,\varphi),(v,\psi)) := \kappa \int_0^1 u\,v \qquad \forall\,((u,\varphi),(v,\psi))\in Q\times Q\,,$$

$F : H \to \mathbb{R}$ is the null functional, and $G : Q \to \mathbb{R}$ is given by

$$G(v,\psi) := -\int_0^1 f\,v - \psi\cdot(a,b) \qquad \forall\,(v,\psi)\in Q\,.$$

The structure of problems (1.27) and (1.28), in particular the one arising with $\kappa = 0$, corresponds to the typical form of a mixed variational formulation. The main aspects of the respective abstract theory are reviewed in Chap. 2. Furthermore, we remark that the idea of introducing Lagrange multipliers to deal with essential boundary conditions will also be used in subsequent chapters when analyzing more complex boundary value problems in two and three dimensions.

1.2.2 A Model in $\mathbb{R}^n$

Let Ω be a bounded domain of $\mathbb{R}^n$, $n \geq 2$, with Lipschitz-continuous boundary Γ. Then, given $f \in L^2(\Omega)$ and $g \in H^{1/2}(\Gamma)$ (see definition of $H^{1/2}(\Gamma)$ and further details in Sect. 1.3.2), we consider the Poisson problem

$$-\,\Delta u = f \quad \text{in} \quad \Omega\,, \qquad u = g \quad \text{on} \quad \Gamma\,. \tag{1.29}$$

The primal formulation of (1.29), which is derived using one of the Green identities (cf. Corollary 1.2 or Theorem 1.8), reduces to the following: find $u \in H^1(\Omega)$ such that $u = g$ on Γ and

$$\int_\Omega \nabla u\cdot\nabla v = \int_\Omega f\,v \qquad \forall\,v \in H_0^1(\Omega) \tag{1.30}$$

[see the beginning of Sect. 1.3 for the definitions of $H^1(\Omega)$ and $H_0^1(\Omega)$ in this n-dimensional case]. Similarly to the analysis for (1.25), we remark here that the Dirichlet boundary condition is not natural but only essential for a primal formulation of (1.30). However, in what follows we show that it becomes natural when the corresponding mixed formulation is utilized instead. In fact, defining the additional unknown $\sigma = \nabla u$ in Ω, problem (1.29) is rewritten as the first-order system

$$\sigma = \nabla u \quad \text{in} \quad \Omega\,, \quad \operatorname{div}(\sigma) = -f \quad \text{in} \quad \Omega\,, \quad u = g \quad \text{in} \quad \Gamma\,.$$

Then, multiplying the equation $\sigma = \nabla u$ in Ω by $\tau \in H(\operatorname{div};\Omega)$, integrating by parts, and using the Dirichlet boundary condition for u, we arrive at

$$\int_{\Omega} \sigma \cdot \tau + \int_{\Omega} u \operatorname{div}(\tau) = \langle \tau \cdot \mathbf{n}, g \rangle \qquad \forall \tau \in H(\operatorname{div};\Omega), \tag{1.31}$$

where $\mathbf{n}$ is the normal vector exterior to Γ and $\langle \cdot,\cdot \rangle$ denotes the duality between $H^{-1/2}(\Gamma)$ and $H^{1/2}(\Gamma)$ (see definition of this duality in Sect. 1.3.4). Recall (which will be utilized subsequently in Sect. 1.3.4) that

$$H(\operatorname{div};\Omega) := \left\{ \tau \in [L^2(\Omega)]^n : \quad \operatorname{div}(\tau) \in L^2(\Omega) \right\}, \tag{1.32}$$

where $\operatorname{div}(\tau) \in L^2(\Omega)$ is meant in the distributional sense, that is, that there exists $z \in L^2(\Omega)$ such that

$$-\int_{\Omega} \nabla\varphi \cdot \tau = \int_{\Omega} z\varphi \qquad \forall \varphi \in C_0^{\infty}(\Omega).$$

Furthermore, it is easy to show, using that $L^2(\Omega)$ is Hilbert, that $H(\operatorname{div};\Omega)$, endowed with the inner product

$$\langle \sigma, \tau \rangle_{\operatorname{div},\Omega} := \int_{\Omega} \left\{ \sigma \cdot \tau + \operatorname{div}(\sigma) \operatorname{div}(\tau) \right\} \qquad \forall \sigma,\, \tau \in H(\operatorname{div};\Omega)$$

and induced norm $\| \cdot \|_{\operatorname{div},\Omega}$, is also Hilbert. As stated earlier, further details on this space, including the proof of the integration by parts formula yielding (1.31), are presented in Sect. 1.3.4.

On the other hand, multiplying the equation $\operatorname{div}(\sigma) = -f$ in Ω by $v \in L^2(\Omega)$, we obtain

$$\int_{\Omega} v \operatorname{div}(\sigma) = -\int_{\Omega} f v \qquad \forall v \in L^2(\Omega). \tag{1.33}$$

Therefore, the mixed variational formulation of (1.29) is obtained by gathering (1.31) and (1.33), which leads to the following problem: find $(\sigma, u) \in H \times Q$ such that

$$\begin{aligned} a(\sigma,\tau) + b(\tau,u) &= \langle \tau \cdot \mathbf{n}, g \rangle \quad \forall \tau \in H, \\ b(\sigma,v) \qquad\qquad &= -\int_{\Omega} f v \quad \forall v \in Q, \end{aligned} \tag{1.34}$$

where $H := H(\operatorname{div};\Omega)$, $Q := L^2(\Omega)$, and $a : H \times H \to \mathbb{R}$, $b : H \times Q \to \mathbb{R}$, are the bilinear forms defined by

$$a(\sigma,\tau) := \int_{\Omega} \sigma \cdot \tau \qquad \forall (\sigma,\tau) \in H \times H,$$

$$b(\tau,v) := \int_{\Omega} v \operatorname{div}(\tau) \qquad \forall (\tau,v) \in H \times Q.$$

As was mentioned in relation to problems (1.27) and (1.28), the structure of (1.34) also corresponds to the typical form of a mixed variational formulation (see the corresponding details in Chap. 2).

As a final comment, and according to what has been observed with the examples of this section, we remark that Dirichlet and Neumann boundary conditions change roles (natural versus essential) when primal and mixed variational formulations are employed. The following table summarizes this fact:

FORMULATION $\longrightarrow$	PRIMAL	MIXED
Dirichlet condition	Essential	Natural
Neumann condition	Natural	Essential

1.3 Traces and Green's Identities

In this section we present some results on traces, integration by parts formulae, and Green's identities for some Sobolev spaces, and particularly for $H(\mathrm{div};\Omega)$. In what follows, given a bounded domain Ω of $\mathbb{R}^n$ with Lipschitz-continuous boundary Γ, the Sobolev space of order 1 is defined as

$$H^1(\Omega) := \left\{ v \in L^2(\Omega) : \quad \frac{\partial v}{\partial x_i} \in L^2(\Omega) \quad \forall i \in \{1,2,\ldots,n\} \right\},$$

where $\dfrac{\partial v}{\partial x_i} \in L^2(\Omega)$ is meant in the distributional sense, that is, that there exists $z_i \in L^2(\Omega)$ such that

$$-\int_\Omega v \frac{\partial \varphi}{\partial x_i} = \int_\Omega z_i \varphi \qquad \forall \varphi \in C_0^\infty(\Omega).$$

It is easy to show, using that $L^2(\Omega)$ is Hilbert, that $H^1(\Omega)$ endowed with the inner product

$$\langle v, w \rangle_{1,\Omega} := \int_\Omega \left\{ \nabla v \cdot \nabla w + vw \right\} \quad \forall v, w \in H^1(\Omega)$$

is also Hilbert. The induced seminorm and norm are given, respectively, by

$$|v|_{1,\Omega} := \|\nabla v\|_{0,\Omega} \quad \text{and} \quad \|v\|_{1,\Omega} := \left\{ |v|_{1,\Omega}^2 + \|v\|_{0,\Omega}^2 \right\}^{1/2} \quad \forall v \in H^1(\Omega).$$

We also define here the closed subspace of $H^1(\Omega)$ given by

$$H_0^1(\Omega) := \overline{C_0^\infty(\Omega)}^{\|\cdot\|_{1,\Omega}},$$

that is, the closure of $C_0^\infty(\Omega)$ in $H^1(\Omega)$ with respect to the norm $\|\cdot\|_{1,\Omega}$. Next, in what follows, we let $\mathscr{D}(\bar{\Omega})$ [or $C_0^\infty(\bar{\Omega})$] be the space of restrictions to Ω of functions that are of class C_0^∞ in an open set containing $\bar{\Omega}$.

1.3.1 Traces of $H^1(\Omega)$

The classical trace theorem in $H^1(\Omega)$ requires the following previous result (cf. [51, Lemme 1.3-5]).

Theorem 1.4 (Trace Inequality). *Let Ω be a bounded domain of $\mathbb{R}^n$ with Lipschitz-continuous boundary Γ, and let $\gamma_0 : \mathscr{D}(\bar{\Omega}) \to L^2(\Gamma)$ be the mapping defined by*

$$\gamma_0(\varphi) := \varphi|_\Gamma \qquad \forall\, \varphi \in \mathscr{D}(\bar{\Omega}).$$

Then there exists $C > 0$ such that

$$\|\gamma_0(\varphi)\|_{0,\Gamma} \leq C\|\varphi\|_{1,\Omega} \qquad \forall\, \varphi \in \mathscr{D}(\bar{\Omega}). \tag{1.35}$$

Theorem 1.5 (Trace Theorem in $H^1(\Omega)$). *Let Ω be a bounded domain of $\mathbb{R}^n$ with Lipschitz-continuous boundary Γ. Then the mapping $\gamma_0 : \mathscr{D}(\bar{\Omega}) \to L^2(\Gamma)$ can be extended by continuity and density to a linear and bounded operator $\gamma_0 : H^1(\Omega) \to L^2(\Gamma)$ such that $\gamma_0(\varphi) := \varphi|_\Gamma \quad \forall\, \varphi \in \mathscr{D}(\bar{\Omega})$.*

Proof. We use the trace inequality (cf. Theorem 1.4) and the fact that $\mathscr{D}(\bar{\Omega})$ is dense in $H^1(\Omega)$ with respect to the norm $\|\cdot\|_{1,\Omega}$. In fact, given $v \in H^1(\Omega)$, we consider a sequence $\{\varphi_j\}_{j\in\mathbb{N}} \subseteq \mathscr{D}(\bar{\Omega})$ such that

$$\|\varphi_j - v\|_{1,\Omega} \to 0 \quad \text{when} \quad j \to +\infty.$$

Then, using the linearity of γ_0 and applying the trace inequality (1.35) to $(\varphi_j - \varphi_k)$, $j, k \in \mathbb{N}$, we obtain

$$\|\gamma_0(\varphi_j) - \gamma_0(\varphi_k)\|_{0,\Gamma} = \|\gamma_0(\varphi_j - \varphi_k)\|_{0,\Gamma} \leq C\|\varphi_j - \varphi_k\|_{1,\Omega},$$

which says that $\{\gamma_0(\varphi_j)\}_{j\in\mathbb{N}}$ is a Cauchy sequence in $L^2(\Gamma)$. Thus, since $L^2(\Gamma)$ is complete, there exists $\xi \in L^2(\Gamma)$ such that

$$\|\gamma_0(\varphi_j) - \xi\|_{0,\Gamma} \to 0 \quad \text{when} \quad j \to +\infty,$$

which suggests setting $\gamma_0(v) := \xi$. However, we must make sure that this is well defined in the sense that the resulting ξ is independent of the chosen sequence. Indeed, let us consider another sequence $\{\tilde{\varphi}_j\}_{j\in\mathbb{N}} \subseteq \mathscr{D}(\bar{\Omega})$ such that

$$\|\tilde{\varphi}_j - v\|_{1,\Omega} \to 0 \quad \text{when} \quad j \to +\infty.$$

It follows that

$$\|\gamma_0(\tilde{\varphi}_j) - \gamma_0(\varphi_j)\|_{0,\Gamma} = \|\gamma_0(\tilde{\varphi}_j - \varphi_j)\|_{0,\Gamma} \le C\|\tilde{\varphi}_j - \varphi_j\|_{1,\Omega} \to 0 \text{ when } j \to +\infty,$$

which shows that $\{\gamma_0(\tilde{\varphi}_j)\}_{j\in\mathbb{N}}$ and $\{\gamma_0(\varphi_j)\}_{j\in\mathbb{N}}$ have the same limit ξ in $L^2(\Gamma)$ as $j \to +\infty$. Therefore, γ_0 is well defined in $H^1(\Omega)$. Moreover, the same argument showing that $\gamma_0(v)$ does not depend on the sequence can be used to see that γ_0 is linear.

On the other hand, taking limit when $j \to +\infty$ in the inequality

$$\|\gamma_0(\varphi_j)\|_{0,\Gamma} \le C\|\varphi_j\|_{1,\Omega},$$

we obtain

$$\|\xi\|_{0,\Gamma} \le C\|v\|_{1,\Omega},$$

which proves that γ_0 is bounded. □

1.3.2 The Space $H^{1/2}(\Gamma)$

We now let $H^{1/2}(\Gamma)$ be the trace space on the boundary Γ, that is,

$$H^{1/2}(\Gamma) := \gamma_0(H^1(\Omega)),$$

which is endowed with the norm

$$\|\xi\|_{1/2,\Gamma} := \inf\left\{ \|v\|_{1,\Omega} : \quad v \in H^1(\Omega) \text{ such that } \gamma_0(v) = \xi \right\} \quad \forall \xi \in H^{1/2}(\Gamma).$$

It is straightforward from the preceding definition that

$$\|\gamma_0(w)\|_{1/2,\Gamma} \le \|w\|_{1,\Omega} \qquad \forall w \in H^1(\Omega),$$

which shows that the operator $\gamma_0 : H^1(\Omega) \to H^{1/2}(\Gamma)$, besides being linear and surjective, is bounded. Then, it can be proved that $N(\gamma_0)$, the null space of γ_0, is given by $H_0^1(\Omega)$ (cf. [51, Théorème 1.4-1]). Equivalently, $v \in N(\gamma_0) = H_0^1(\Omega)$ if and only if $v \in H^1(\Omega)$ and there exists a sequence $\{\varphi_j\}_{j\in\mathbb{N}} \subseteq C_0^\infty(\Omega)$ such that $\|\varphi_j - v\|_{1,\Omega} \to 0$ when $j \to +\infty$.

We thus conclude that $\tilde{\gamma}_0 := \gamma_0|_{H_0^1(\Omega)^\perp} : H_0^1(\Omega)^\perp \to H^{1/2}(\Gamma)$ is a linear and bounded bijection. It follows that $\tilde{\gamma}_0^{-1} : H^{1/2}(\Gamma) \to H_0^1(\Omega)^\perp$ is a linear bijection as well, and that $\tilde{\gamma}_0^{-1}$ is a right inverse of γ_0, that is,

$$\gamma_0 \tilde{\gamma}_0^{-1} = I : H^{1/2}(\Gamma) \to H^{1/2}(\Gamma).$$

In addition, the following lemma implies the boundedness of $\tilde{\gamma}_0^{-1}$, which in turn allows us to conclude that $(H^{1/2}(\Gamma), \|\cdot\|_{1/2,\Gamma})$ is complete.

Lemma 1.3. *There holds*

$$\|\xi\|_{1/2,\Gamma} := \|\tilde{\gamma}_0^{-1}(\xi)\|_{1,\Omega} \qquad \forall \xi \in H^{1/2}(\Gamma). \tag{1.36}$$

Proof. Given $v \in H^1(\Omega)$, we consider its unique decomposition $v = v_0 + v_0^\perp$, with $v_0 \in H_0^1(\Omega)$ and $v_0^\perp \in H_0^1(\Omega)^\perp$. Then, letting $\gamma_0(v) = \xi$, it follows that $\xi = \gamma_0(v_0^\perp)$, which yields $v_0^\perp = \tilde{\gamma}_0^{-1}(\xi)$. According to the foregoing analysis, we have

$$\inf\left\{ \|v\|_{1,\Omega} : \quad v \in H^1(\Omega) \text{ such that } \gamma_0(v) = \xi \right\}$$
$$= \inf_{v_0 \in H_0^1(\Omega)} \|v_0 + \tilde{\gamma}_0^{-1}(\xi)\|_{1,\Omega},$$

and since $\tilde{\gamma}_0^{-1}(\xi) \in H_0^1(\Omega)^\perp$, the infimum is clearly attained at $v_0 = 0$, which yields (1.36). □

Corollary 1.1. $(H^{1/2}(\Gamma), \|\cdot\|_{1/2,\Gamma})$ *is complete.*

Proof. It is a direct consequence of (1.36) and the fact that $(H_0^1(\Omega)^\perp, \|\cdot\|_{1,\Omega})$, being a closed subspace of $H^1(\Omega)$, is also Hilbert. □

1.3.3 Integration by Parts and Green's Identities

The density of $\mathscr{D}(\bar{\Omega})$ in $H^1(\Omega)$ and the trace theorem (cf. Theorem 1.5) allow us to prove the following result.

Theorem 1.6 (Integration by Parts Formula). *Let Ω be a bounded domain of $\mathbb{R}^n$ with Lipschitz-continuous boundary Γ. Then for each $v, w \in H^1(\Omega)$ there holds*

$$\int_\Omega v \frac{\partial w}{\partial x_i} = -\int_\Omega w \frac{\partial v}{\partial x_i} + \int_\Gamma \gamma_0(v)\,\gamma_0(w)\,n_i \qquad \forall i \in \{1,2,\ldots,n\}, \tag{1.37}$$

where n_i is the ith component of the normal vector $\mathbf{n}$.

Proof. Let $v, w \in H^1(\Omega)$, and let us consider the sequences $\{\varphi_j\}_{j\in\mathbb{N}}, \{\psi_j\}_{j\in\mathbb{N}} \subseteq \mathscr{D}(\bar{\Omega})$ such that

$$\|\varphi_j - v\|_{1,\Omega} \to 0 \quad \text{and} \quad \|\psi_j - w\|_{1,\Omega} \to 0 \quad \text{when} \quad j \to +\infty. \tag{1.38}$$

According to the trace theorem, we know that

$$\|\gamma_0(\varphi_j) - \gamma_0(v)\|_{0,\Gamma} \to 0 \quad \text{and} \quad \|\gamma_0(\psi_j) - \gamma_0(w)\|_{0,\Gamma} \to 0 \quad \text{when} \quad j \to +\infty. \tag{1.39}$$

Now, applying the classical integration by parts formula for smooth functions (which is a consequence of Gauss'divergence theorem), we obtain

$$\int_\Omega \varphi_j \frac{\partial \psi_j}{\partial x_i} = -\int_\Omega \psi_j \frac{\partial \varphi_j}{\partial x_i} + \int_\Gamma \gamma_0(\varphi_j)\,\gamma_0(\psi_j)\,n_i \qquad \forall i \in \{1,2,\ldots,n\}. \quad (1.40)$$

Then, taking limit in (1.40) when $j \to +\infty$ we get (1.37). In fact, adding and subtracting a convenient term, applying the Cauchy–Schwarz inequality, and noting that the sequence $\left\{ \frac{\partial \psi_j}{\partial x_i} \right\}_{j\in\mathbb{N}}$ is bounded in $L^2(\Omega)$ [because $\{\psi_j\}_{j\in\mathbb{N}}$ is convergent in $H^1(\Omega)$], we deduce that

$$\begin{aligned}\left| \int_\Omega \varphi_j \frac{\partial \psi_j}{\partial x_i} - \int_\Omega v \frac{\partial w}{\partial x_i} \right| &\le \left| \int_\Omega (\varphi_j - v) \frac{\partial \psi_j}{\partial x_i} \right| + \left| \int_\Omega v \left(\frac{\partial \psi_j}{\partial x_i} - \frac{\partial w}{\partial x_i} \right) \right| \\ &\le \|\varphi_j - v\|_{0,\Omega} \left\| \frac{\partial \psi_j}{\partial x_i} \right\|_{0,\Omega} + \|v\|_{0,\Omega} \left\| \frac{\partial \psi_j}{\partial x_i} - \frac{\partial w}{\partial x_i} \right\|_{0,\Omega} \\ &\le C \|\varphi_j - v\|_{0,\Omega} + \|v\|_{0,\Omega} |\psi_j - w|_{1,\Omega},\end{aligned}$$

from which, thanks to (1.38), we obtain

$$\lim_{j\to+\infty} \int_\Omega \varphi_j \frac{\partial \psi_j}{\partial x_i} = \int_\Omega v \frac{\partial w}{\partial x_i}.$$

Analogously, one can prove that

$$\lim_{j\to+\infty} \int_\Omega \psi_j \frac{\partial \varphi_j}{\partial x_i} = \int_\Omega w \frac{\partial v}{\partial x_i}.$$

Finally, similar arguments, but this time using (1.39), yield

$$\lim_{j\to+\infty} \int_\Gamma \gamma_0(\varphi_j)\,\gamma_0(\psi_j)\,n_i = \int_\Gamma \gamma_0(v)\,\gamma_0(w)\,n_i,$$

which completes the proof. □

An immediate consequence of the preceding theorem is given by the following corollary, which requires the definition of the Sobolev space of order 2:

$$H^2(\Omega) := \left\{ v \in H^1(\Omega) : \quad \frac{\partial^2 v}{\partial x_i x_j} \in L^2(\Omega) \quad \forall i, j \in \{1,2,\ldots,n\} \right\}. \quad (1.41)$$

Corollary 1.2 (Green's Identity). *Let Ω be a bounded domain of $\mathbb{R}^n$ with Lipschitz-continuous boundary Γ. Then for each $v \in H^1(\Omega)$ and $u \in H^2(\Omega)$ there holds*

$$\int_\Omega v\,\Delta u = -\int_\Omega \nabla u \cdot \nabla v + \int_\Gamma \gamma_0(v)\,\gamma_0(\nabla u)\cdot \mathbf{n}, \quad (1.42)$$

where $\gamma_0(\nabla u)$ is the vector arising from a componentwise application of γ_0 to ∇u.

Proof. Let $v \in H^1(\Omega)$ and $u \in H^2(\Omega)$. It is clear from (1.41) that $\nabla u \in [H^1(\Omega)]^n$. Then, applying (1.37) with v and $w := \frac{\partial u}{\partial x_i}$, both in $H^1(\Omega)$, we obtain

$$\int_\Omega v \frac{\partial^2 u}{\partial x_i^2} = -\int_\Omega \frac{\partial u}{\partial x_i}\frac{\partial v}{\partial x_i} + \int_\Gamma \gamma_0(v)\,\gamma_0\left(\frac{\partial u}{\partial x_i}\right) n_i \qquad \forall i \in \{1,2,\ldots,n\},$$

from which, summing with respect to i, we arrive at (1.42). □

It is important to remark here that the primal formulation (1.30) of the Poisson problem (1.29) is derived precisely by applying the identity (1.42) and then using that $\gamma_0(v) = 0$ for each $v \in H_0^1(\Omega)$. Furthermore, the expression $\gamma_0(\nabla u)\cdot \mathbf{n}$ is usually written $\frac{\partial u}{\partial \mathbf{n}}$, and it is called the normal derivative of u.

1.3.4 Normal Traces of $H(\mathrm{div};\Omega)$

In this section we prove that the vector functions of $H(\mathrm{div};\Omega)$ [cf. (1.32)] have normal traces on the boundary Γ. To this end, in what follows we denote by $H^{-1/2}(\Gamma)$ the dual of $H^{1/2}(\Gamma)$. In addition, given a functional $\psi \in H^{-1/2}(\Gamma)$, its evaluation in $\xi \in H^{1/2}(\Gamma)$ is usually written as $\langle \psi, \xi \rangle$, that is,

$$\langle \psi, \xi \rangle := \psi(\xi) \qquad \forall \xi \in H^{1/2}(\Gamma),$$

which is why, as stated in the preface and in Sect. 1.2.2, $\langle \cdot, \cdot \rangle$ is also called the duality between $H^{-1/2}(\Gamma)$ and $H^{1/2}(\Gamma)$.

Theorem 1.7 (Normal Trace of $H(\mathrm{div};\Omega)$). *Let Ω be a bounded domain of $\mathbb{R}^n$ with Lipschitz-continuous boundary Γ. Then there exists a linear, bounded, and surjective operator $\gamma_{\mathbf{n}} : H(\mathrm{div};\Omega) \to H^{-1/2}(\Gamma)$ such that for each $\tau \in [H^1(\Omega)]^n$, $\gamma_{\mathbf{n}}(\tau)$ is identified, through the inner product of $L^2(\Gamma)$, with $\gamma_0(\tau)\cdot \mathbf{n}$.*

Proof. Let $\tilde{\gamma}_0^{-1} : H^{1/2}(\Gamma) \to H_0^1(\Omega)^\perp$ be the right inverse of γ_0. Then, given $\tau \in H(\mathrm{div};\Omega)$, we define the linear functional $\gamma_{\mathbf{n}}(\tau) : H^{1/2}(\Gamma) \to \mathbb{R}$ as

$$\gamma_{\mathbf{n}}(\tau)(\xi) := \int_\Omega \tau \cdot \nabla \tilde{\gamma}_0^{-1}(\xi) + \int_\Omega \tilde{\gamma}_0^{-1}(\xi)\,\mathrm{div}(\tau) \qquad \forall \xi \in H^{1/2}(\Gamma). \tag{1.43}$$

Note that the linearity of $\gamma_{\mathbf{n}}(\tau)$ follows from that of $\tilde{\gamma}_0^{-1}$. In addition, applying the Cauchy–Schwarz inequality, we obtain that

$$|\gamma_{\mathbf{n}}(\tau)(\xi)| \le \|\tau\|_{0,\Omega}\,\|\nabla \tilde{\gamma}_0^{-1}(\xi)\|_{0,\Omega} + \|\tilde{\gamma}_0^{-1}(\xi)\|_{0,\Omega}\,\|\mathrm{div}(\tau)\|_{0,\Omega}$$

$$\le \|\tau\|_{\mathrm{div},\Omega}\,\|\tilde{\gamma}_0^{-1}(\xi)\|_{1,\Omega} = \|\tau\|_{\mathrm{div},\Omega}\,\|\xi\|_{1/2,\Gamma} \quad \forall \xi \in H^{1/2}(\Gamma),$$

which shows that $\gamma_{\mathbf{n}}(\tau)$ is bounded, and therefore it belongs to $H^{-1/2}(\Gamma)$, with

$$\|\gamma_{\mathbf{n}}(\tau)\|_{-1/2,\Gamma} \leq \|\tau\|_{\mathrm{div},\Omega}\,. \tag{1.44}$$

The preceding analysis supports the definition of the linear and bounded operator

$$\begin{aligned} \gamma_{\mathbf{n}} : H(\mathrm{div};\Omega) &\to H^{-1/2}(\Gamma) \\ \tau \quad &\to \gamma_{\mathbf{n}}(\tau)\,, \end{aligned}$$

which, in virtue of (1.44), satisfies $\|\gamma_{\mathbf{n}}\| \leq 1$. Now, if we consider the vector function $\tau := (\tau_1, \tau_2, \ldots, \tau_n)^{\mathrm{t}} \in [H^1(\Omega)]^n$, the integration by parts formula (1.37) gives

$$\begin{aligned} \int_\Omega \tilde{\gamma}_0^{-1}(\xi)\,\mathrm{div}(\tau) &= \sum_{i=1}^n \int_\Omega \tilde{\gamma}_0^{-1}(\xi)\frac{\partial \tau_i}{\partial x_i} \\ &= \sum_{i=1}^n \left\{ -\int_\Omega \tau_i \frac{\partial}{\partial x_i}\tilde{\gamma}_0^{-1}(\xi) + \int_\Gamma \xi\,\gamma_0(\tau_i)\,n_i \right\} \\ &= -\int_\Omega \tau\cdot\nabla\tilde{\gamma}_0^{-1}(\xi) + \int_\Gamma \xi\,\gamma_0(\tau)\cdot\mathbf{n}, \end{aligned}$$

and hence, whenever $\tau \in [H^1(\Omega)]^n$, (1.43) reduces to

$$\langle \gamma_{\mathbf{n}}(\tau), \xi\rangle = \int_\Gamma \gamma_0(\tau)\cdot\mathbf{n}\,\xi = \langle \gamma_0(\tau)\cdot\mathbf{n}, \xi\rangle_{0,\Gamma} \qquad \forall\,\xi \in H^{1/2}(\Gamma), \tag{1.45}$$

where $\langle\cdot,\cdot\rangle_{0,\Gamma}$ is the usual inner product of $L^2(\Gamma)$.

It remains to show the surjectivity of the operator $\gamma_{\mathbf{n}}$. In other words, given $\psi \in H^{-1/2}(\Gamma) := (H^{1/2}(\Gamma))'$, we must prove that there exists $\tau \in H(\mathrm{div};\Omega)$ such that $\psi = \gamma_{\mathbf{n}}(\tau)$. In fact, let us define the subspace

$$\tilde{H}^1(\Omega) := \left\{ v \in H^1(\Omega) : \quad \int_\Omega v = 0 \right\},$$

and let us consider the following variational problem: find $z \in \tilde{H}^1(\Omega)$ such that

$$\int_\Omega \nabla z\cdot\nabla w = F(w) \qquad \forall\,w \in H^1(\Omega), \tag{1.46}$$

where

$$F(w) := -\frac{\langle \psi, \gamma_0(1)\rangle}{|\Omega|}\int_\Omega w + \langle \psi, \gamma_0(w)\rangle \quad \forall\,w \in H^1(\Omega). \tag{1.47}$$

Note that the linearity of the integral, ψ, and γ_0 guarantees that F is linear. In addition, it is clear that F is also bounded since

$$|F(w)| \leq \frac{|\langle \psi, \gamma_0(1)\rangle|}{|\Omega|^{1/2}}\,\|w\|_{0,\Omega} + \|\psi\|_{-1/2,\Gamma}\,\|w\|_{1,\Omega} \leq C\|w\|_{1,\Omega} \quad \forall\,w \in H^1(\Omega).$$

Now, utilizing the decomposition $H^1(\Omega) = \tilde{H}^1(\Omega) \oplus \mathbb{R}$ and the fact that $F(1) = 0$, it is easily proved that (1.46) is equivalent to the following problem: find $z \in \tilde{H}^1(\Omega)$ such that

$$\int_\Omega \nabla z \cdot \nabla w = F(w) \qquad \forall w \in \tilde{H}^1(\Omega). \tag{1.48}$$

Thus, since the norm $\|\cdot\|_{1,\Omega}$ and seminorm $|\cdot|_{1,\Omega}$ of $H^1(\Omega)$ are equivalent in $\tilde{H}^1(\Omega)$, which follows from the generalized Poincaré inequality (cf. [46, Theorem 5.11.2]), we can apply the Lax–Milgram lemma to conclude that (1.48) [and hence (1.46)] has a unique solution $z \in \tilde{H}^1(\Omega)$. Then we define $\tau := \nabla z$ in Ω, which certainly belongs to $[L^2(\Omega)]^n$. However, it follows from (1.46) and (1.47) that

$$\int_\Omega \tau \cdot \nabla w = -\frac{\langle \psi, \gamma_0(1)\rangle}{|\Omega|} \int_\Omega w \qquad \forall w \in C_0^\infty(\Omega),$$

which says, in the distributional sense, that $\operatorname{div}(\tau) = \dfrac{\langle \psi, \gamma_0(1)\rangle}{|\Omega|}$ in Ω, and hence $\tau \in H(\operatorname{div};\Omega)$. In this way, the formulation (1.46) can be rewritten as

$$\int_\Omega \tau \cdot \nabla w + \int_\Omega w \operatorname{div}(\tau) = \langle \psi, \gamma_0(w)\rangle \qquad \forall w \in H^1(\Omega),$$

so that, taking in particular $w = \tilde{\gamma}_0^{-1}(\xi)$, with $\xi \in H^{1/2}(\Gamma)$, we find that

$$\langle \psi, \xi\rangle = \int_\Omega \tau \cdot \nabla \tilde{\gamma}_0^{-1}(\xi) + \int_\Omega \tilde{\gamma}_0^{-1}(\xi) \operatorname{div}(\tau) = \langle \gamma_{\mathbf{n}}(\tau), \xi\rangle \qquad \forall \xi \in H^{1/2}(\Gamma),$$

which shows that $\psi = \gamma_{\mathbf{n}}(\tau)$. □

We prove next that when $\tau \in [H^1(\Omega)]^n$, the functional $\gamma_{\mathbf{n}}(\tau)$ can be defined not only as given in (1.43), but also by employing any $w \in H^1(\Omega)$ such that $\gamma_0(w) = \xi \in H^{1/2}(\Gamma)$. In fact, in this case we know from (1.45) that

$$\langle \gamma_{\mathbf{n}}(\tau), \xi\rangle := \langle \gamma_0(\tau) \cdot \mathbf{n}, \xi\rangle_{0,\Gamma} = \int_\Gamma \gamma_0(\tau) \cdot \mathbf{n}\, \gamma_0(w) = \sum_{i=1}^n \int_\Gamma \gamma_0(w)\, \gamma_0(\tau_i)\, n_i,$$

which, applying the integration by parts formula (1.37), yields

$$\langle \gamma_{\mathbf{n}}(\tau), \xi\rangle := \langle \gamma_0(\tau) \cdot \mathbf{n}, \xi\rangle_{0,\Gamma} = \int_\Omega \tau \cdot \nabla w + \int_\Omega w \operatorname{div}(\tau).$$

Equivalently, the preceding analysis is summarized in the following identity:

$$\langle \gamma_{\mathbf{n}}(\tau), \gamma_0(w)\rangle := \langle \gamma_0(\tau) \cdot \mathbf{n}, \gamma_0(w)\rangle_{0,\Gamma} = \int_\Omega \tau \cdot \nabla w + \int_\Omega w \operatorname{div}(\tau) \tag{1.49}$$
$$\forall w \in H^1(\Omega), \quad \forall \tau \in [H^1(\Omega)]^n.$$

Moreover, the following lemma utilizes the density of $[C_0^\infty(\bar{\Omega})]^n$ in $H(\mathrm{div};\Omega)$ to extend (1.49) to the case of $\tau \in H(\mathrm{div};\Omega)$.

Lemma 1.4 (Green's Identity in $H(\mathrm{div};\Omega)$). *Let Ω be a bounded domain of $\mathbb{R}^n$ with Lipschitz-continuous boundary Γ. Then there holds*

$$\langle \gamma_{\mathbf{n}}(\tau), \gamma_0(w)\rangle = \int_\Omega \tau \cdot \nabla w + \int_\Omega w \,\mathrm{div}(\tau) \quad \forall w \in H^1(\Omega), \quad \forall \tau \in H(\mathrm{div};\Omega). \tag{1.50}$$

Proof. Let $w \in H^1(\Omega)$ and $\tau \in H(\mathrm{div};\Omega)$. Since $[C_0^\infty(\bar{\Omega})]^n$ is dense in $H(\mathrm{div};\Omega)$, there exists a sequence $\{\mathbf{z}_k\}_{k\in\mathrm{N}} \subseteq [C_0^\infty(\bar{\Omega})]^n$ such that

$$\lim_{k\to+\infty} \|\mathbf{z}_k - \tau\|_{\mathrm{div},\Omega} = 0. \tag{1.51}$$

Then, applying (1.49) to w and $\mathbf{z}_k$ we obtain

$$\langle \gamma_{\mathbf{n}}(\mathbf{z}_k), \gamma_0(w)\rangle := \langle \gamma_0(\mathbf{z}_k)\cdot \mathbf{n}, \gamma_0(w)\rangle_{0,\Gamma} = \int_\Omega \mathbf{z}_k \cdot \nabla w + \int_\Omega w\,\mathrm{div}(\mathbf{z}_k) \qquad \forall k \in \mathbb{N},$$

whence, taking limit when $k \to +\infty$, and employing the continuity of $\gamma_{\mathbf{n}}$ [cf. (1.44)] and the convergence (1.51), we conclude (1.50). □

We find it important to remark here that when $\tau \in H(\mathrm{div};\Omega)$, the evaluation $\langle \gamma_{\mathbf{n}}(\tau), \gamma_0(w)\rangle$ cannot be replaced by the expression $\langle \gamma_0(\tau)\cdot \mathbf{n}, \gamma_0(w)\rangle_{0,\Gamma}$ since the latter only makes sense if $\tau \in [H^1(\Omega)]^n$. However, from the preceding proof we know that for each sequence $\{\mathbf{z}_k\}_{k\in\mathrm{N}} \subseteq [C_0^\infty(\bar{\Omega})]^n$ converging to $\tau \in H(\mathrm{div};\Omega)$, we can write

$$\langle \gamma_{\mathbf{n}}(\tau), \gamma_0(w)\rangle = \lim_{k\to+\infty} \langle \gamma_0(\mathbf{z}_k)\cdot \mathbf{n}, \gamma_0(w)\rangle_{0,\Gamma} \qquad \forall w \in H^1(\Omega),$$

which is why it is usually said that $\langle \cdot,\cdot\rangle$ denotes the duality between $H^{-1/2}(\Gamma)$ and $H^{1/2}(\Gamma)$ with respect to the inner product $\langle \cdot,\cdot\rangle_{0,\Gamma}$ of $L^2(\Gamma)$.

On the other hand, it is also interesting to mention that the first equation of the primal formulation (1.34), that is, (1.31), follows precisely from the Green identity (1.50).

We end this chapter with the following theorem establishing an interesting consequence of the preceding results.

Theorem 1.8. *Let Ω be a bounded domain of $\mathbb{R}^n$ with Lipschitz-continuous boundary Γ, and let us define $H^1_\Delta(\Omega) := \left\{ v \in H^1(\Omega) : \quad \Delta v \in L^2(\Omega) \right\}$. Then there exists a linear and bounded operator $\gamma_1 : H^1_\Delta(\Omega) \to H^{-1/2}(\Gamma)$ such that for each $u \in H^2(\Omega)$, $\gamma_1(u)$ is identified, by means of the inner product of $L^2(\Gamma)$, with $\gamma_0(\nabla u)\cdot \mathbf{n}$, that is,*

$$\langle \gamma_1(u), \xi\rangle = \langle \gamma_0(\nabla u)\cdot \mathbf{n}, \xi\rangle_{0,\Gamma} \qquad \forall \xi \in H^{1/2}(\Gamma), \quad \forall u \in H^2(\Omega).$$

Moreover, there holds

$$\langle \gamma_1(u), \gamma_0(w) \rangle = \int_\Omega \nabla u \cdot \nabla w + \int_\Omega w \Delta u \quad \forall w \in H^1(\Omega), \quad \forall u \in H^1_\Delta(\Omega). \quad (1.52)$$

Proof. Since $\nabla u \in H(\mathrm{div};\Omega)$ for each $u \in H^1_\Delta(\Omega)$, it suffices to set $\gamma_1 := \gamma_{\mathbf{n}} \circ \nabla$, that is,

$$\gamma_1(u) := \gamma_{\mathbf{n}}(\nabla u) \qquad \forall u \in H^1_\Delta(\Omega),$$

and then apply Theorem 1.7 and Lemma 1.4. □

Note that the preceding theorem significantly improves the Green identity (1.42) given in Corollary 1.2 since the operator γ_1 makes it possible to extend the notion of normal derivative to the whole space $H^1_\Delta(\Omega)$, which is strictly larger than $H^2(\Omega)$.

Chapter 2
Babuška–Brezzi Theory

In this chapter we present the main results forming part of the Babuška–Brezzi theory, which makes it possible to analyze a large family of mixed variational formulations and their respective Galerkin approximations. Our main references here include [16, 41, 50, 52]. We begin by introducing the specific kind of operator equations that we are interested in.

2.1 Operator Equation

Let $(H, \langle\cdot,\cdot\rangle_H)$ and $(Q, \langle\cdot,\cdot\rangle_Q)$ be real Hilbert spaces with induced norms $\|\cdot\|_H$ and $\|\cdot\|_Q$, respectively, and let $a : H \times H \to \mathbb{R}$ and $b : H \times Q \to \mathbb{R}$ be bounded bilinear forms. Then, given $F \in H'$ and $G \in Q'$, we are interested in the following problem: find $(\sigma, u) \in H \times Q$ such that

$$\begin{aligned} a(\sigma,\tau) + b(\tau,u) &= F(\tau) \qquad \forall\, \tau \in H\,, \\ b(\sigma,v) \qquad\qquad &= G(v) \qquad \forall\, v \in Q\,. \end{aligned} \tag{2.1}$$

Next, let $\mathbf{A} : H \to H$ and $\mathbf{B} : H \to Q$ be the linear and bounded operators induced by a and b, respectively. Equivalently, according to the analysis in Sect. 1.1 [cf. (1.2)–(1.4)], there holds

$$\mathbf{A} := \mathscr{R}_H \circ \mathscr{A} \quad \text{and} \quad \mathbf{B} := \mathscr{R}_Q \circ \mathscr{B}\,,$$

where $\mathscr{R}_H : H' \to H$ and $\mathscr{R}_Q : Q' \to Q$ are the respective Riesz mappings, and the operators $\mathscr{A} : H \to H'$ and $\mathscr{B} : H \to Q'$ are defined by

$$\mathscr{A}(\sigma)(\tau) := a(\sigma,\tau) \qquad \forall\, \sigma \in H\,, \quad \forall\, \tau \in H$$

and

$$\mathscr{B}(\tau)(v) := b(\tau,v) \qquad \forall\, \tau \in H\,, \quad \forall\, v \in Q\,.$$

G.N. Gatica, *A Simple Introduction to the Mixed Finite Element Method: Theory and Applications*, SpringerBriefs in Mathematics, DOI 10.1007/978-3-319-03695-3_2,

It follows that

$$a(\sigma,\tau) = \langle \mathbf{A}(\sigma),\tau\rangle_H \qquad \forall(\sigma,\tau)\in H\times H \tag{2.2}$$

and

$$b(\tau,v) = \langle \mathbf{B}(\tau),v\rangle_Q = \langle \mathbf{B}^*(v),\tau\rangle_H \qquad \forall(\tau,v)\in H\times Q, \tag{2.3}$$

where $\mathbf{B}^* : Q \to H$ is the adjoint operator of $\mathbf{B}$.

In this way, (2.1) is rewritten, equivalently, as follows: find $(\sigma,u)\in H\times Q$ such that

$$\begin{aligned} \langle \mathbf{A}(\sigma),\tau\rangle_H + \langle \mathbf{B}^*(u),\tau\rangle_H &= \langle \mathscr{R}_H(F),\tau\rangle_H \qquad \forall \tau\in H,\\ \langle \mathbf{B}(\sigma),v\rangle_Q \qquad\qquad\qquad &= \langle \mathscr{R}_Q(G),v\rangle_Q \qquad \forall v\in Q, \end{aligned}$$

or: find $(\sigma,u)\in H\times Q$ such that

$$\begin{aligned} \mathbf{A}(\sigma) + \mathbf{B}^*(u) &= \mathscr{R}_H(F),\\ \mathbf{B}(\sigma) \qquad\quad &= \mathscr{R}_Q(G), \end{aligned} \tag{2.4}$$

which, denoting the null operator by 0, reduces to the following matrix operator equation: find $(\sigma,u)\in H\times Q$ such that

$$\begin{pmatrix} \mathbf{A} & \mathbf{B}^* \\ \mathbf{B} & 0 \end{pmatrix} \begin{pmatrix} \sigma \\ u \end{pmatrix} = \begin{pmatrix} \mathscr{R}_H(F) \\ \mathscr{R}_Q(G) \end{pmatrix}. \tag{2.5}$$

We now aim to provide the conditions that are necessary and sufficient for (2.1) [equivalently (2.4) or (2.5)] to be well-posed.

2.2 The inf-sup Condition

We recall first that this condition was already introduced in Sect. 1.1 [cf. (1.16)–(1.19)]. Indeed, we say that the bounded bilinear form $b : H\times Q\to\mathbb{R}$ satisfies the continuous inf-sup condition if there exists a constant $\beta>0$ such that

$$\sup_{\substack{\tau\in H\\ \tau\neq 0}} \frac{b(\tau,v)}{\|\tau\|_H} \geq \beta\,\|v\|_Q \qquad \forall v\in Q. \tag{2.6}$$

Note, as was established for the pairs of conditions (1.16)–(1.18) and (1.17)–(1.19), (2.6) is equivalent to

$$\inf_{\substack{v\in Q\\ v\neq 0}} \sup_{\substack{\tau\in H\\ \tau\neq 0}} \frac{b(\tau,v)}{\|\tau\|_H\,\|v\|_Q} \geq \beta,$$

which explains again the name INF-SUP. This hypothesis is also known as the LADYZHENSKAYA–BABUŠKA–BREZZI condition, or simply BABUŠKA–BREZZI condition. In addition, utilizing the adjoint operator $\mathbf{B}^*$, we have

$$\sup_{\substack{\tau\in H\\ \tau\neq 0}} \frac{b(\tau,v)}{\|\tau\|_H} = \sup_{\substack{\tau\in H\\ \tau\neq 0}} \frac{\langle \mathbf{B}^*(v),\tau\rangle_H}{\|\tau\|_H} = \|\mathbf{B}^*(v)\|_H,$$

and therefore condition (2.6) is written also as

$$\|\mathbf{B}^*(v)\|_H \geq \beta\,\|v\|_Q \qquad \forall v\in Q. \tag{2.7}$$

Moreover, the following lemma establishes equivalent conditions for (2.6) [or (2.7)].

Lemma 2.1. *The following statements are equivalent:*

(i) *There exists $\beta>0$ such that*

$$\sup_{\substack{\tau\in H\\ \tau\neq 0}} \frac{b(\tau,v)}{\|\tau\|_H} \geq \beta\,\|v\|_Q \qquad \forall v\in Q.$$

(ii) $\mathbf{B}^*$ *is an isomorphism (linear bijection) from Q into $N(\mathbf{B})^\perp$, and*

$$\|\mathbf{B}^*(v)\|_H \geq \beta\,\|v\|_Q \qquad \forall v\in Q.$$

(iii) $\mathbf{B}$ *is an isomorphism (linear bijection) from $N(\mathbf{B})^\perp$ into Q, and*

$$\|\mathbf{B}(\tau)\|_Q \geq \beta\,\|\tau\|_H \qquad \forall \tau\in N(\mathbf{B})^\perp. \tag{2.8}$$

(iv) $\mathbf{B}: H\to Q$ *is surjective.*

Proof.
(i) $\Rightarrow$ (ii): Suppose that there exists $\beta>0$ such that (2.6) [or, equivalently, (2.7)] is satisfied. It follows from (2.7) that $N(\mathbf{B}^*)=\{0\}$ and $R(\mathbf{B}^*)$ is closed, which implies that $\mathbf{B}^*$ is injective and $R(\mathbf{B}^*)=N((\mathbf{B}^*)^*)^\perp=N(\mathbf{B})^\perp$. Thus, $\mathbf{B}^*$ is a linear bijection from Q into $N(\mathbf{B})^\perp$.
(ii) $\Rightarrow$ (iii): Suppose that $\mathbf{B}^*$ is a linear bijection from Q into $N(\mathbf{B})^\perp$ and that (2.7) is satisfied. It follows again from (2.7) that $N(\mathbf{B}^*)=\{0\}$ and $R(\mathbf{B}^*)$ is closed. The latter implies, in virtue of a known result from functional analysis, that $R(\mathbf{B})$ is also closed, and therefore $R(\mathbf{B})=N(\mathbf{B}^*)^\perp=\{0\}^\perp=Q$. In this way, $\mathbf{B}$ is a linear bijection from $N(\mathbf{B})^\perp$ into Q. In addition, it is clear from (2.7) that $\|(\mathbf{B}^*)^{-1}\|\leq\frac{1}{\beta}$, and hence

$$\|\mathbf{B}^{-1}\| = \|(\mathbf{B}^{-1})^*\| = \|(\mathbf{B}^*)^{-1}\| \leq \frac{1}{\beta},$$

which yields (2.8).

(iii) $\Rightarrow$ (iv): This follows directly from the fact that $\mathbf{B} : N(\mathbf{B})^{\perp} \to Q$ is bijective (in particular, surjective) and that $H = N(\mathbf{B}) \oplus N(\mathbf{B})^{\perp}$.
(iv) $\Rightarrow$ (i): Suppose now that $\mathbf{B} : H \to Q$ is surjective. Since $R(\mathbf{B}) = Q$ is obviously closed, we have that $R(\mathbf{B}^*)$ is closed as well. Furthermore, applying orthogonality to the identity $Q = R(\mathbf{B}) = N(\mathbf{B}^*)^{\perp}$, we obtain that $N(\mathbf{B}^*) = \{0\}$, which says that $\mathbf{B}^*$ is injective. Hence, the characterization result for operators with a closed range implies inequality (2.7), which is exactly (i).

□

2.3 Main Result

The characterization of the inf-sup condition given by Lemma 2.1 is essential for the proof of the following theorem, which establishes sufficient conditions for (2.1) to be well-posed. In the statement of this theorem and in what follows throughout the rest of this section, we assume the same notations and definitions from the previous sections.

Theorem 2.1. *Let $V := N(\mathbf{B})$ and let $\Pi : H \to V$ be the orthogonal projection operator. Assume that:*

(i) *$\Pi\mathbf{A} : V \to V$ is a bijection;*
(ii) *The bilinear form b satisfies the inf-sup condition* (2.6) [*equivalently,* (2.7)].

Then for each pair $(F, G) \in H' \times Q'$ there exists a unique $(\sigma, u) \in H \times Q$ solution of (2.1) [*equivalently,* (2.4) *or* (2.5)]. *Moreover, there exists a constant $C > 0$, which depends on $\|\mathbf{A}\|$, $\|(\Pi\mathbf{A})^{-1}\|$, and β, such that*

$$\|(\sigma, u)\|_{H\times Q} \le C\Big\{ \|F\|_{H'} + \|G\|_{Q'} \Big\}. \tag{2.9}$$

Proof. Since $\mathbf{B}$ is a bijection from $V^{\perp}$ into Q [which is a consequence of (2.6) and Lemma 2.1, part (iii)], we deduce that there exists a unique $\sigma_g \in V^{\perp}$ such that

$$\mathbf{B}(\sigma_g) = \mathscr{R}_Q(G), \tag{2.10}$$

and according to (2.8), there holds

$$\|\sigma_g\|_H \le \frac{1}{\beta}\|\mathbf{B}(\sigma_g)\|_Q = \frac{1}{\beta}\|G\|_{Q'}. \tag{2.11}$$

Next, since $\Pi\mathbf{A} : V \to V$ is a bijection and $\Pi(\mathscr{R}_H(F) - \mathbf{A}(\sigma_g))$ belongs to V, there exists a unique $\sigma_0 \in V$ such that $\Pi\mathbf{A}(\sigma_0) = \Pi(\mathscr{R}_H(F) - \mathbf{A}(\sigma_g))$. In addition, the bounded inverse theorem guarantees the existence of $\tilde{C} := \|(\Pi\mathbf{A})^{-1}\|$ such that

$$\|\sigma_0\|_H \le \tilde{C}\|\Pi(\mathscr{R}_H(F) - \mathbf{A}(\sigma_g))\|_H \le \tilde{C}\|\mathscr{R}_H(F) - \mathbf{A}(\sigma_g)\|_H,$$

from which, using the bound for $\|\sigma_g\|_H$, we obtain that

$$\|\sigma_0\|_H \leq \tilde{C} \left\{ \|F\|_{H'} + \frac{1}{\beta} \|\mathbf{A}\| \, \|G\|_{Q'} \right\}. \tag{2.12}$$

Now, thanks to the orthogonality condition of the projector Π, it is easy to see that the identity $\Pi\mathbf{A}(\sigma_0) = \Pi(\mathscr{R}_H(F) - \mathbf{A}(\sigma_g))$ is equivalent to saying that the vector $\mathbf{A}(\sigma_0 + \sigma_g) - \mathscr{R}_H(F)$ belongs to $V^\perp$. Thus, it follows from Lemma 2.1, part (ii), that there exists a unique $u \in Q$ such that

$$\mathbf{B}^*(u) = \mathscr{R}_H(F) - \mathbf{A}(\sigma_0 + \sigma_g) \tag{2.13}$$

and

$$\|u\|_Q \leq \frac{1}{\beta} \|\mathbf{B}^*(u)\|_H = \frac{1}{\beta} \|\mathscr{R}_H(F) - \mathbf{A}(\sigma_0 + \sigma_g)\|_H ,$$

whence

$$\|u\|_Q \leq \frac{1}{\beta} \left\{ \|F\|_{H'} + \|\mathbf{A}\| \left(\|\sigma_0\|_H + \|\sigma_g\|_H \right) \right\}. \tag{2.14}$$

In this way, defining $\sigma := \sigma_0 + \sigma_g \in H$ and noting that $\mathbf{B}(\sigma_0) = 0$, we deduce from (2.10), (2.13), and the estimates (2.11), (2.12), and (2.14), that (σ, u) solves (2.4) and satisfies (2.9).

For the uniqueness, let $(\sigma, u) \in H \times Q$ be a solution of the homogeneous problem

$$\begin{aligned} \mathbf{A}(\sigma) + \mathbf{B}^*(u) &= 0, \\ \mathbf{B}(\sigma) \qquad\quad &= 0. \end{aligned}$$

It is clear from the second equation that $\sigma \in V$, and then, applying the projector Π to the first one, and recalling that $\mathbf{B}^*(u) \in V^\perp$, we obtain $\Pi\mathbf{A}(\sigma) = 0$. Thus, since $\Pi\mathbf{A} : V \to V$ is a bijection, it follows that $\sigma = 0$, and then from the first equation we obtain that $\mathbf{B}^*(u) = 0$. Finally, since $\mathbf{B}^* : Q \to V^\perp$ is also a bijection, we conclude that $u = 0$.

□

We show next that conditions (i) and (ii) of Theorem 2.1 are also **necessary**. Indeed, we have the following result.

Theorem 2.2. *Let $V := N(\mathbf{B})$, and let $\Pi : H \to V$ be the orthogonal projection operator. Assume that for each pair $(F, G) \in H' \times Q'$ there exists a unique solution $(\sigma, u) \in H \times Q$ of (2.1) [equivalently, (2.4) or (2.5)] that satisfies*

$$\|(\sigma, u)\|_{H \times Q} \leq C \left\{ \|F\|_{H'} + \|G\|_{Q'} \right\},$$

with a constant $C > 0$ independent of F and G. Then:

(i) $\Pi \mathbf{A} : V \to V$ *is a bijection;*
(ii) *The bilinear form* b *satisfies the inf-sup condition* (2.6).

Proof. First we prove (ii). For this purpose, in virtue of Lemma 2.1, it suffices to show that $\mathbf{B}$ is surjective. In fact, given $g \in Q$, we know from the hypotheses that there exists a unique pair $(\sigma_g, u_g) \in H \times Q$ such that

$$\begin{aligned} \mathbf{A}(\sigma_g) + \mathbf{B}^*(u_g) &= 0, \\ \mathbf{B}(\sigma_g) \qquad\quad &= g, \end{aligned}$$

and it is clear that the second equation of this system confirms the surjectivity of $\mathbf{B}$. Hence, knowing that b satisfies the continuous inf-sup condition, we can use the equivalences given by Lemma 2.1 to show that $\Pi \mathbf{A} : V \to V$ is a bijection.

Indeed, given $f \in V$, we know also from the hypotheses that there exists a unique $(\sigma_f, u_f) \in H \times Q$ such that

$$\begin{aligned} \mathbf{A}(\sigma_f) + \mathbf{B}^*(u_f) &= f, \\ \mathbf{B}(\sigma_f) \qquad\quad &= 0, \end{aligned}$$

which shows, according to the second equation, that $\sigma_f \in V$. Then, applying the orthogonal projector Π to the first equation, and using, from part (ii) of Lemma 2.1, that $\mathbf{B}^*(u_f) \in V^\perp$, we obtain $\Pi \mathbf{A}(\sigma_f) = \Pi(f) = f$, thereby proving that $\Pi \mathbf{A} : V \to V$ is surjective. Then, let $\sigma_0 \in V$ be such that $\Pi \mathbf{A}(\sigma_0) = 0$. It follows that $\mathbf{A}(\sigma_0) \in V^\perp$, and since, according to part (ii) of Lemma 2.1, $\mathbf{B}^* : Q \to V^\perp$ is a bijection, we deduce that there exists a unique $u_0 \in Q$ such that $\mathbf{B}^*(u_0) = -\mathbf{A}(\sigma_0)$. In this way we have

$$\begin{aligned} \mathbf{A}(\sigma_0) + \mathbf{B}^*(u_0) &= 0, \\ \mathbf{B}(\sigma_0) \qquad\quad &= 0, \end{aligned}$$

and, thanks again to the hypotheses, we obtain $(\sigma_0, u_0) = (0,0)$, which gives the injectivity of $\Pi \mathbf{A} : V \to V$.

□

On the other hand, it is easy to see, according to Lemma 1.2, inequalities (1.16) and (1.17), and the orthogonality characterizing the projector $\Pi : H \to V$, that hypothesis (i) in Theorems 2.1 and 2.2 is equivalent to each one of the following pairs of conditions:

(i-1) There exists $\alpha > 0$ such that

$$\sup_{\substack{\tau \in V \\ \tau \neq 0}} \frac{a(\sigma, \tau)}{\|\tau\|_H} \geq \alpha \|\sigma\|_H \qquad \forall \sigma \in V;$$

(i-2) For each $\tau \in V$, $\tau \neq 0$, there holds $\quad \sup_{\sigma \in V} a(\sigma, \tau) > 0$;

and

(i-1)$'$ There exists $\alpha > 0$ such that

$$\sup_{\substack{\tau \in V \\ \tau \neq 0}} \frac{a(\tau,\sigma)}{\|\tau\|_H} \geq \alpha \|\sigma\|_H \qquad \forall \sigma \in V;$$

(i-2)$'$ For each $\tau \in V$, $\tau \neq 0$, there holds $\sup_{\sigma \in V} a(\tau,\sigma) > 0$.

More precisely, hypothesis (i-1) (resp. (i-1)$'$) is an inf-sup condition for the bilinear form a, which is the same as requiring that the operator $\Pi\,\mathbf{A}$ [resp. $(\Pi\,\mathbf{A})^*$] be injective and with a closed range, which, in addition, is equivalent to the surjectivity of the operator $(\Pi\,\mathbf{A})^*$ (resp. $\Pi\,\mathbf{A}$). Then, (i-2) [resp. (i-2)$'$] is equivalent to the injectivity of $(\Pi\,\mathbf{A})^*$ (resp. $\Pi\,\mathbf{A}$).

Certainly, when a is a symmetric bilinear form on $V \times V$, the operator $\Pi\,\mathbf{A}$ becomes self-adjoint, and in this case (i-2) and (i-2)$'$ are redundant and therefore unnecessary.

Furthermore, it is important to remark that a sufficient (but not necessary) condition for (i), which appears very often in applications, is the V-ellipticity of the bilinear form a, which means (cf. Definition 1.3) that there exists $\alpha > 0$ such that

$$a(\tau,\tau) \geq \alpha \|\tau\|_H^2 \qquad \forall \tau \in V. \tag{2.15}$$

In fact, the result that usually appears in the literature, even more frequently than Theorem 2.1, is the following.

Theorem 2.3. *Let* $V := N(\mathbf{B})$ *and assume that:*

(i) *The bilinear form a is V-elliptic* [*cf.* (2.15)].
(ii) *The bilinear form b satisfies the inf-sup condition* (2.6) [*equivalently,* (2.7)].

Then for each pair $(F,G) \in H' \times Q'$ *there exists a unique* $(\sigma,u) \in H \times Q$ *solution of* (2.1) [*equivalently* (2.4) *or* (2.5)]. *Moreover, there exists a constant* $C > 0$, *which depends on* $\|\mathbf{A}\|$, α, *and* β, *such that*

$$\|(\sigma,u)\|_{H\times Q} \leq C \left\{ \|F\|_{H'} + \|G\|_{Q'} \right\}.$$

Proof. It suffices to see, for instance in virtue of the Lax–Milgram lemma, that the V-ellipticity of a implies hypothesis (i) of Theorem 2.1. □

Further extensions of the Babuška–Brezzi theory to other classes of linear and nonlinear abstract variational problems have been developed in several works (e.g., [13, 26, 31, 45]). In addition, interesting characterizations of the inf-sup condition for bilinear forms defined on product spaces can be found in [40] and [45].

2.4 Application Examples

In this section we illustrate the applicability of the Babuška–Brezzi theory with the classical examples given by the Poisson and elasticity problems.

2.4.1 Poisson Problem

Let Ω be a bounded domain of $\mathbb{R}^n, n \geq 2$, with Lipschitz-continuous boundary Γ. Then, given $f \in L^2(\Omega)$ and $g \in H^{1/2}(\Gamma)$, we consider the same problem introduced in Sect. 1.2.2, that is,

$$-\Delta u = f \quad \text{in} \quad \Omega, \quad u = g \quad \text{on} \quad \Gamma. \tag{2.16}$$

Then, as in that section, we introduce the additional unknown $\sigma := \nabla u$ in Ω, so that problem (2.16) is rewritten as the first-order system

$$\sigma = \nabla u \quad \text{in} \quad \Omega, \quad \operatorname{div} \sigma = -f \quad \text{in} \quad \Omega, \quad u = g \quad \text{on} \quad \Gamma.$$

Then, multiplying the equation $\sigma = \nabla u$ in Ω by $\tau \in H(\operatorname{div}; \Omega)$, and applying the Green identity (1.50) (cf. Lemma 1.4), we obtain

$$\int_\Omega \sigma \cdot \tau = \int_\Omega \nabla u \cdot \tau = -\int_\Omega u \operatorname{div} \tau + \langle \gamma_{\mathbf{n}}(\tau), \gamma_0(u) \rangle,$$

from which, using that the Dirichlet boundary condition says that $\gamma_0(u) = g$ on Γ, we deduce that

$$\int_\Omega \sigma \cdot \tau + \int_\Omega u \operatorname{div} \tau = \langle \gamma_{\mathbf{n}}(\tau), g \rangle \quad \forall \tau \in H(\operatorname{div}; \Omega). \tag{2.17}$$

Recall here that $\gamma_0 : H^1(\Omega) \to H^{1/2}(\Gamma)$ and $\gamma_{\mathbf{n}} : H(\operatorname{div}; \Omega) \to H^{-1/2}(\Gamma)$ are the trace operators examined in Sects. 1.3.1 and 1.3.4 and that $\langle \cdot, \cdot \rangle$ denotes the duality between $H^{-1/2}(\Gamma)$ and $H^{1/2}(\Gamma)$ with respect to the inner product of $L^2(\Gamma)$. Note also that the Green identity (1.50) and (2.17) justify the integration by parts employed in the deduction of (1.31).

On the other hand, the equilibrium equation $\operatorname{div} \sigma = -f$ in Ω, is rewritten as

$$\int_\Omega v \operatorname{div} \sigma = -\int_\Omega f v \quad \forall v \in L^2(\Omega). \tag{2.18}$$

Consequently, gathering (2.17) and (2.18), we find that the mixed variational formulation of (2.16) reduces to the following: find $(\sigma, u) \in H \times Q$ such that

$$\begin{aligned} a(\sigma, \tau) + b(\tau, u) &= F(\tau) \quad \forall \tau \in H, \\ b(\sigma, v) \qquad &= G(v) \quad \forall v \in Q, \end{aligned} \tag{2.19}$$

where

$$H := H(\mathrm{div};\Omega), \quad Q := L^2(\Omega),$$

a and b are the bilinear forms defined by

$$a(\sigma,\tau) := \int_\Omega \sigma\cdot\tau \qquad \forall\,(\sigma,\tau)\in H\times H,$$

$$b(\tau,v) := \int_\Omega v\,\mathrm{div}\,\tau \qquad \forall\,(\tau,v)\in H\times Q,$$

and the functionals $F\in H'$ and $G\in Q'$ are given by

$$F(\tau) := \langle \gamma_{\mathbf{n}}(\tau), g\rangle \quad \forall\,\tau\in H, \qquad G(v) := -\int_\Omega f\,v \quad \forall\,v\in Q. \tag{2.20}$$

In what follows we apply the particular case of the Babuška–Brezzi theory given by Theorem 2.3. In fact, we first observe that a and b are clearly bounded with

$$\|\mathbf{A}\| \le 1 \quad \text{and} \quad \|\mathbf{B}\| \le 1,$$

where $\mathbf{A}: H\to H$ and $\mathbf{B}: H\to Q$ are the operators induced by a and b, respectively. Next, it is clear that $\mathbf{B}(\tau) := \mathrm{div}\,\tau \quad \forall\,\tau\in H$, and hence

$$V := N(\mathbf{B}) = \Big\{\tau\in H: \quad \mathbf{B}(\tau)=0\Big\} = \Big\{\tau\in H(\mathrm{div};\Omega): \quad \mathrm{div}\,\tau = 0 \quad \text{in} \quad \Omega\Big\}.$$

It follows that

$$a(\tau,\tau) = \|\tau\|_{0,\Omega}^2 = \|\tau\|_{\mathrm{div},\Omega}^2 \qquad \forall\,\tau\in V,$$

which shows that a is V-elliptic with ellipticity constant $\alpha = 1$.

Furthermore, keeping in mind from Lemma 2.1 that the continuous inf-sup condition for b is equivalent to the surjectivity of $\mathbf{B}$, we now let $v\in Q$ and consider the boundary value problem

$$-\Delta z = v \quad \text{in} \quad \Omega, \quad z = 0 \quad \text{on} \quad \Gamma,$$

whose primal variational formulation reads as follows: find $z\in H_0^1(\Omega)$ such that

$$\int_\Omega \nabla z\cdot\nabla w = \int_\Omega v\,w \quad \forall\,w\in H_0^1(\Omega). \tag{2.21}$$

It follows from the Lax–Milgram lemma (cf. Theorem 1.1) that (2.21) has a unique solution $z\in H_0^1(\Omega)$, which satisfies

$$|z|_{1,\Omega} \le \tilde{C}\,\|v\|_{0,\Omega}, \tag{2.22}$$

where $\tilde{C}>0$ is a constant arising from the n-dimensional version of the Friedrichs–Poincaré inequality provided by Lemma 1.1 (cf. [51, Théorème 1.2-5]). Then, defining $\tilde{\tau} := -\nabla z$ in Ω, we obtain $\mathrm{div}\,\tilde{\tau} = v$ in Ω, and thus $\tilde{\tau}\in H(\mathrm{div};\Omega)$, which proves that $\mathbf{B} :=$ div is surjective. On the other hand, thanks to the boundedness of the operator $\gamma_{\mathbf{n}}$ (cf. Theorem 1.7), the duality $\langle \cdot, \cdot \rangle$, and the Cauchy–Schwarz inequality in $L^2(\Omega)$, we easily see from (2.20) that F and G are bounded with

$$\|F\|_{H'} \le \|g\|_{1/2,\Gamma} \quad \text{and} \quad \|G\|_{Q'} \le \|f\|_{0,\Omega}.$$

Therefore, Theorem 2.3 implies that there exists a unique pair $(\sigma, u) \in H \times Q$ solution of the mixed variational formulation (2.19) that satisfies

$$\|(\sigma, u)\|_{H \times Q} \le C \left\{ \|g\|_{1/2,\Gamma} + \|f\|_{0,\Omega} \right\}. \tag{2.23}$$

According to the details shown in the proof of Theorem 2.1, $C > 0$ in (2.23) depends on the constant β for the continuous inf-sup condition of b, and $\|\mathbf{A}\| \le 1$, where $\mathbf{A}$ is the operator induced by the bilinear form a. In this respect, and in order to have a closer idea of the value of β, we observe from the definition of $\tilde{\tau}$, and using the inequality (2.22), that

$$\|\tilde{\tau}\|^2_{\mathrm{div},\Omega} = \|\tilde{\tau}\|^2_{0,\Omega} + \|\mathrm{div}\, \tilde{\tau}\|^2_{0,\Omega} = |z|^2_{1,\Omega} + \|v\|^2_{0,\Omega} \le (1 + \tilde{C}^2)\, \|v\|^2_{0,\Omega}.$$

Thus, it follows that

$$\sup_{\substack{\tau \in H \\ \tau \neq 0}} \frac{b(\tau, v)}{\|\tau\|_H} \ge \frac{b(\tilde{\tau}, v)}{\|\tilde{\tau}\|_H} = \frac{\int_\Omega v \,\mathrm{div}\, \tilde{\tau}}{\|\tilde{\tau}\|_{\mathrm{div},\Omega}} = \frac{\|v\|^2_{0,\Omega}}{\|\tilde{\tau}\|_{\mathrm{div},\Omega}} \ge \frac{1}{(1 + \tilde{C}^2)^{1/2}} \|v\|_{0,\Omega},$$

from which we deduce that one can set $\beta := \dfrac{1}{(1 + \tilde{C}^2)^{1/2}}$.

2.4.2 Poisson Problem with Mixed Boundary Conditions

Let Ω be a bounded domain of $\mathbb{R}^n$, $n \ge 2$, with Lipschitz-continuous boundary Γ, and let Γ_D and Γ_N be disjoint parts of Γ such that $|\Gamma_D| \neq 0$ and $\Gamma = \overline{\Gamma}_D \cup \overline{\Gamma}_N$. Then, given $f \in L^2(\Omega)$ and $g \in H^{-1/2}_{00}(\Gamma_N)$, we are interested in the boundary value problem

$$-\Delta u = f \quad \text{in} \quad \Omega, \qquad u = 0 \quad \text{on} \quad \Gamma_D, \qquad \nabla u \cdot \mathbf{n} = g \quad \text{on} \quad \Gamma_N, \tag{2.24}$$

where $\mathbf{n}$ is the normal vector to Γ. We recall here that $H^{-1/2}_{00}(\Gamma_N)$ is the dual of $H^{1/2}_{00}(\Gamma_N)$, where

$$H^{1/2}_{00}(\Gamma_N) := \left\{ v|_{\Gamma_N} : \quad v \in H^1(\Omega), \quad v = 0 \text{ on } \Gamma_D \right\}.$$

Equivalently, if $E_{N,0} : H^{1/2}(\Gamma_N) \to L^2(\Gamma)$ is the extension operator

$$E_{N,0}(\eta) := \begin{cases} \eta & \text{on} \quad \Gamma_N \\ 0 & \text{on} \quad \Gamma_D \end{cases} \qquad \forall\, \eta \in H^{1/2}(\Gamma_N),$$

then

$$H_{00}^{1/2}(\Gamma_N) := \Big\{ \eta \in H^{1/2}(\Gamma_N) : \quad E_{N,0}(\eta) \in H^{1/2}(\Gamma) \Big\},$$

which is endowed with the norm

$$\|\eta\|_{1/2,00,\Gamma_N} := \|E_{N,0}(\eta)\|_{1/2,\Gamma} \qquad \forall\, \eta \in H_{00}^{1/2}(\Gamma_N). \tag{2.25}$$

Then, the duality between $H_{00}^{-1/2}(\Gamma_N)$ and $H_{00}^{1/2}(\Gamma_N)$ is denoted by $\langle \cdot, \cdot \rangle_{\Gamma_N}$. In addition, given $\psi \in H^{-1/2}(\Gamma)$, its restriction to Γ_N, denoted by $\psi|_{\Gamma_N}$ and defined by

$$\langle \psi|_{\Gamma_N}, \eta \rangle_{\Gamma_N} := \langle \psi, E_{N,0}(\eta) \rangle \quad \forall\, \eta \in H_{00}^{1/2}(\Gamma_N), \tag{2.26}$$

where $\langle \cdot, \cdot \rangle$ denotes the duality between $H^{-1/2}(\Gamma)$ and $H^{1/2}(\Gamma)$, clearly belongs to $H_{00}^{-1/2}(\Gamma_N)$. Moreover, it is clear from (2.25) and (2.26) that

$$\|\psi|_{\Gamma_N}\|_{-1/2,00,\Gamma_N} := \sup_{\substack{\eta \in H_{00}^{1/2}(\Gamma_N) \\ \eta \neq 0}} \frac{\langle \psi|_{\Gamma_N}, \eta \rangle_{\Gamma_N}}{\|\eta\|_{1/2,00,\Gamma_N}} = \sup_{\substack{\eta \in H_{00}^{1/2}(\Gamma_N) \\ \eta \neq 0}} \frac{\langle \psi, E_{N,0}(\eta) \rangle}{\|E_{N,0}(\eta)\|_{1/2,\Gamma}},$$

which yields

$$\|\psi|_{\Gamma_N}\|_{-1/2,00,\Gamma_N} \le \|\psi\|_{-1/2,\Gamma} \qquad \forall\, \psi \in H^{-1/2}(\Gamma). \tag{2.27}$$

Now, for the mixed variational formulation of (2.24) we proceed analogously to the previous example and define the additional unknown $\sigma := \nabla u$ in Ω, so that (2.24) is rewritten as

$$\sigma = \nabla u \quad \text{in} \quad \Omega, \quad \operatorname{div} \sigma = -f \quad \text{in} \quad \Omega,$$
$$u = 0 \quad \text{on} \quad \Gamma_D, \quad \sigma \cdot \mathbf{n} = g \quad \text{on} \quad \Gamma_N.$$

Then, applying again the Green identity (1.50) (cf. Lemma 1.4), we obtain

$$\int_\Omega \sigma \cdot \tau = \int_\Omega \nabla u \cdot \tau = -\int_\Omega u \operatorname{div} \tau + \langle \gamma_{\mathbf{n}}(\tau), \gamma_0(u) \rangle,$$

and introducing the auxiliary unknown $\xi := -\gamma_0(u) \in H_{00}^{1/2}(\Gamma_N)$, which is supported by the Dirichlet boundary condition, we arrive at

$$\int_\Omega \sigma \cdot \tau + \int_\Omega u \operatorname{div} \tau + \langle \gamma_{\mathbf{n}}(\tau)|_{\Gamma_N}, \xi \rangle_{\Gamma_N} = 0 \qquad \forall\, \tau \in H(\operatorname{div}; \Omega). \tag{2.28}$$

On the other hand, as in the previous section, the equation $\operatorname{div}\sigma = -f$ in Ω is weakly imposed as

$$\int_\Omega v \operatorname{div}\sigma = -\int_\Omega f v \quad \forall \in L^2(\Omega). \tag{2.29}$$

Finally, since $\gamma_{\mathbf{n}}(\sigma)|_{\Gamma_N} \in H_{00}^{-1/2}(\Gamma_N)$, the Neumann boundary condition $\sigma \cdot \mathbf{n} = g$ on Γ_N is equivalently reformulated as

$$\langle \gamma_{\mathbf{n}}(\sigma)|_{\Gamma_N}, \eta \rangle_{\Gamma_N} = \langle g, \eta \rangle_{\Gamma_N} \quad \forall \eta \in H_{00}^{1/2}(\Gamma_N). \tag{2.30}$$

Consequently, by gathering (2.29) and (2.30) into a single equation and placing it together with (2.28) we deduce that the mixed variational formulation of (2.24) reduces to the following: find $(\sigma, (u, \xi)) \in H \times Q$ such that

$$\begin{aligned} a(\sigma, \tau) + b(\tau, (u, \xi)) &= F(\tau) \quad \forall \tau \in H, \\ b(\sigma, (v, \eta)) \qquad\qquad &= G(v, \eta) \quad \forall (v, \eta) \in Q, \end{aligned} \tag{2.31}$$

where

$$H := H(\operatorname{div}; \Omega), \quad Q = L^2(\Omega) \times H_{00}^{1/2}(\Gamma_N),$$

a is the same bilinear form of the previous example (cf. Sect. 2.4.1), $b : H \times Q \to \mathbb{R}$ is defined by

$$b(\tau, (v, \eta)) = \int_\Omega v \operatorname{div}\tau + \langle \gamma_{\mathbf{n}}(\tau)|_{\Gamma_N}, \eta \rangle_{\Gamma_N} \quad \forall (\tau, (v, \eta)) \in H \times Q,$$

and the functionals $F \in H'$ and $G \in Q'$ are given by

$$F(\tau) := 0 \quad \forall \tau \in H, \quad G(v, \eta) := -\int_\Omega f v + \langle g, \eta \rangle_{\Gamma_N} \quad \forall (v, \eta) \in Q.$$

In what follows we again apply Theorem 2.3. First of all, we observe that $\mathbf{B} : H \to Q$, the operator induced by b, is given by

$$\mathbf{B}(\tau) := (\operatorname{div}\tau, \mathscr{R}_{00}\, \gamma_{\mathbf{n}}(\tau)|_{\Gamma_N}) \quad \forall \tau \in H, \tag{2.32}$$

where $\mathscr{R}_{00} : H_{00}^{-1/2}(\Gamma_N) \to H_{00}^{1/2}(\Gamma_N)$ is the corresponding Riesz mapping. We recall here that $H_{00}^{1/2}(\Gamma_N)$ is a Hilbert space with the inner product

$$\langle \chi, \eta \rangle_{1/2,00,\Gamma_N} := \langle E_{N,0}(\chi), E_{N,0}(\eta) \rangle_{1/2,\Gamma} \qquad \forall \chi, \eta \in H_{00}^{1/2}(\Gamma_N),$$

where $\langle \cdot, \cdot \rangle_{1/2,\Gamma}$ is the inner product of $H^{1/2}(\Gamma)$. Since $\mathscr{R}_{00}$ is an isometry and $\gamma_{\mathbf{n}}|_{\Gamma_N}$ [cf. (2.26)] is bounded [cf. (1.44) in Theorem 1.7], it follows that $\mathbf{B}$ is also bounded with $\|\mathbf{B}\| \le 2$. Then, it is clear that

$$V := \mathbf{N}(\mathbf{B}) = \left\{ \tau \in H : \operatorname{div}\tau = 0 \quad \text{in} \quad \Omega, \quad \gamma_{\mathbf{n}}(\tau) = 0 \quad \text{on} \quad \Gamma_N \right\},$$

and hence, for each $\tau \in V$ there holds

$$a(\tau,\tau) = \|\tau\|_{0,\Omega}^2 = \|\tau\|_{\mathrm{div},\Omega}^2 \,,$$

which proves that a is V-elliptic with ellipticity constant $\alpha = 1$.

On the other hand, given $(v,\eta) \in Q$, we consider the boundary value problem

$$\Delta z = v \quad \text{in} \quad \Omega, \quad z = 0 \quad \text{on} \quad \Gamma_D, \quad \nabla z \cdot \mathbf{n} = \mathscr{R}_{00}^{-1}(\eta) \quad \text{on} \quad \Gamma_N,$$

whose primal variational formulation reads as follows: find $z \in H^1_{\Gamma_D}(\Omega)$ such that

$$\int_\Omega \nabla z \cdot \nabla w = -\int_\Omega v w + \langle \mathscr{R}_{00}^{-1}(\eta), \gamma_0(w)\rangle_{\Gamma_N} \quad \forall w \in H^1_{\Gamma_D}(\Omega), \tag{2.33}$$

where

$$H^1_{\Gamma_D}(\Omega) := \Big\{ w \in H^1(\Omega) : \quad \gamma_0(w) = 0 \quad \text{on} \quad \Gamma_D \Big\}.$$

The generalized Poincaré inequality (cf. [46, Theorem 5.11.2]) yields the equivalence between $\|\cdot\|_{1,\Omega}$ and $|\cdot|_{1,\Omega}$ in $H^1_{\Gamma_D}(\Omega)$, and hence the Lax–Milgram lemma (cf. Theorem 1.1) implies that (2.33) has a unique solution $z \in H^1_{\Gamma_D}(\Omega)$. It follows, according to the respective continuous dependence result and the duality between $H_{00}^{-1/2}(\Gamma_N)$ and $H_{00}^{1/2}(\Gamma_N)$, that

$$|z|_{1,\Omega} \le \hat{C}\Big\{ \|v\|_{0,\Omega} + \|\eta\|_{1/2,00,\Gamma_N} \Big\}, \tag{2.34}$$

where $\hat{C} > 0$ is a constant arising from that inequality and the trace inequality (cf. Theorem 1.4). Thus, defining $\hat{\tau} := \nabla z$ in Ω, we have that $\operatorname{div} \hat{\tau} = v$ in Ω, which yields $\hat{\tau} \in H(\mathrm{div};\Omega)$. In addition, there holds $\gamma_{\mathbf{n}}(\hat{\tau})|_{\Gamma_N} = \mathscr{R}_{00}^{-1}(\eta)$ on Γ_N, from which it is clear that $\mathscr{R}_{00}\,\gamma_{\mathbf{n}}(\hat{\tau})|_{\Gamma_N} = \eta$ on Γ_N. This shows, due to (2.32), that $\mathbf{B}(\hat{\tau}) = (v,\eta)$, and therefore $\mathbf{B}$ is surjective. Finally, it is easy to see that G is bounded with

$$\|G\|_{Q'} \le \Big\{ \|f\|_{0,\Omega} + \|g\|_{-1/2,00,\Gamma_N} \Big\},$$

where $\|\cdot\|_{-1/2,00,\Gamma_N}$ stands for the norm of $H_{00}^{-1/2}(\Gamma_N)$. In this way, a direct application of Theorem 2.3 implies that there exists a unique solution $(\sigma,(u,\xi)) \in H \times Q$ of problem (2.31) that satisfies

$$\|(\sigma,(u,\xi))\|_{H\times Q} \le C\Big\{ \|f\|_{0,\Omega} + \|g\|_{-1/2,00,\Gamma_N} \Big\}.$$

Similarly as in the previous example, we remark here that C is a positive constant depending on the constant β for the continuous inf-sup condition of b, and $\|\mathbf{A}\| = 1$, where $\mathbf{A}$ is the operator induced by a. Then, to have a closer idea of the value of β, we observe first from the definition of $\hat{\tau}$, and using (2.34), that

$$\|\hat{\tau}\|_{\mathrm{div},\Omega}^2 = |z|_{1,\Omega}^2 + \|v\|_{0,\Omega}^2 \le \left(1 + 2\hat{C}^2\right) \|(v,\eta)\|_Q^2.$$

Employing this last inequality and denoting by $\langle\cdot,\cdot\rangle_Q$ the inner product of Q, we find that

$$\sup_{\substack{\tau\in H\\ \tau\neq 0}}\frac{b(\tau,(v,\eta))}{\|\tau\|_H}\geq\frac{b(\hat{\tau},(v,\eta))}{\|\hat{\tau}\|_H}=\frac{\displaystyle\int_\Omega v\,\mathrm{div}\,\hat{\tau}+\langle\gamma_{\mathbf{n}}(\hat{\tau}),\eta\rangle_{\Gamma_N}}{\|\hat{\tau}\|_{\mathrm{div},\Omega}}$$

$$=\frac{\langle\mathbf{B}(\hat{\tau}),(v,\eta)\rangle_Q}{\|\hat{\tau}\|_{\mathrm{div},\Omega}}=\frac{\|(v,\eta)\|_Q^2}{\|\hat{\tau}\|_{\mathrm{div},\Omega}}$$

$$\geq\frac{1}{(1+2\hat{C}^2)^{1/2}}\|(v,\eta)\|_Q,$$

from which we conclude that we can set $\beta:=\dfrac{1}{(1+2\hat{C}^2)^{1/2}}$.

2.4.3 Linear Elasticity Problem

Before defining the problem of interest we need to introduce some further notations. In what follows, given a normed space H and $n\in\{2,3\}$, we denote by $\mathbf{H}$ and $\mathbb{H}$ the spaces H^n and $H^{n\times n}$, respectively. In particular, given a domain $\mathscr{O}$, a Lipschitz-continuous curve Σ, and $r\in\mathbb{R}$, we set

$$\mathbf{H}^r(\mathscr{O}):=[H^r(\mathscr{O})]^n,\quad \mathbb{H}^r(\mathscr{O}):=[H^r(\mathscr{O})]^{n\times n},\text{ and }\mathbf{H}^r(\Sigma):=[H^r(\Sigma)]^n,$$

with corresponding norms $\|\cdot\|_{r,\mathscr{O}}$ [for $H^r(\mathscr{O})$, $\mathbf{H}^r(\mathscr{O})$ and $\mathbb{H}^r(\mathscr{O})$] and $\|\cdot\|_{r,\Sigma}$ [for $H^r(\Sigma)$ and $\mathbf{H}^r(\Sigma)$]. Then, when $r=0$, we usually set $\mathbf{L}^2(\mathscr{O})$, $\mathbb{L}^2(\mathscr{O})$, and $\mathbf{L}^2(\Sigma)$ instead of $\mathbf{H}^0(\mathscr{O})$, $\mathbb{H}^0(\mathscr{O})$, and $\mathbf{H}^0(\Sigma)$, respectively. In addition, denoting by $\mathbf{div}$ the usual divergence operator div acting along each row of the tensors, we also let

$$\mathbb{H}(\mathbf{div};\Omega):=\left\{\tau\in\mathbb{L}^2(\Omega):\quad \mathbf{div}\,\tau\in\mathbf{L}^2(\Omega)\right\},$$

which is endowed with the norm $\|\tau\|_{\mathbf{div},\Omega}:=\left\{\|\tau\|_{0,\Omega}^2+\|\mathbf{div}\,\tau\|_{0,\Omega}^2\right\}^{1/2}$.

Now, let Ω be a bounded and simply connected domain of $\mathbb{R}^n$, $n\in\{2,3\}$, with Lipschitz-continuous boundary Γ, and let Γ_D and Γ_N be disjoint parts of Γ such that $|\Gamma_D|\neq 0$ and $\Gamma=\overline{\Gamma}_D\cup\overline{\Gamma}_N$. Then the linear elasticity problem consists in determining the displacement $\mathbf{u}$ and the stress tensor σ of an elastic material occupying the region Ω. More precisely, given $\mathbf{f}\in\mathbf{L}^2(\Omega)$ and $\mathbf{g}\in\mathbf{H}_{00}^{-1/2}(\Gamma_N)$, we look for a symmetric tensor σ and a vector field $\mathbf{u}$ such that

$$\begin{aligned}\sigma=\mathscr{C}\,\mathbf{e}(\mathbf{u})\quad\text{in}\quad\Omega,\qquad \mathbf{div}\,\sigma=-\mathbf{f}\quad\text{in}\quad\Omega,\\ \mathbf{u}=0\quad\text{on}\quad\Gamma_D,\qquad \sigma\mathbf{n}=\mathbf{g}\quad\text{on}\quad\Gamma_N,\end{aligned}\tag{2.35}$$

where $\mathbf{e}(\mathbf{u}) := \frac{1}{2}(\nabla\mathbf{u} + (\nabla\mathbf{u})^{\mathrm{t}})$ is the strain tensor (or symmetric part of $\nabla\mathbf{u}$), $\mathbf{n}$ is the normal vector to Γ, and $\mathscr{C}$ is the elasticity operator given by Hooke's law, that is,

$$\mathscr{C}\zeta := \lambda\,\mathrm{tr}(\zeta)\mathbf{I} + 2\mu\,\zeta \quad \forall\zeta \in \mathbb{L}^2(\Omega). \tag{2.36}$$

Here, $\lambda,\,\mu > 0$ are the respective Lamé constants, $\mathbf{I}$ is the identity matrix of $\mathbb{R}^n$, and tr is the usual matrix trace. Applying tr to (2.36) we can invert this law, which gives

$$\mathscr{C}^{-1}\zeta = \frac{1}{2\mu}\,\zeta - \frac{\lambda}{2\mu\,(n\lambda + 2\mu)}\,\mathrm{tr}(\zeta)\mathbf{I} \qquad \forall\zeta \in \mathbb{L}^2(\Omega).$$

2.4.3.1 Dirichlet Boundary Conditions

In this section we consider the particular case of (2.35) given by $\Gamma_D = \Gamma$, that is

$$\sigma = \mathscr{C}\mathbf{e}(\mathbf{u}) \quad \text{in} \quad \Omega, \qquad \mathbf{div}\,\sigma = -\mathbf{f} \quad \text{in} \quad \Omega, \qquad \mathbf{u} = 0 \quad \text{on} \quad \Gamma. \tag{2.37}$$

In order to derive a mixed variational formulation of (2.37) we notice first that

$$\mathscr{C}^{-1}\sigma = \mathbf{e}(\mathbf{u}) = \nabla\mathbf{u} - \rho\,, \tag{2.38}$$

where

$$\rho := \frac{1}{2}(\nabla\mathbf{u} - (\nabla\mathbf{u})^{\mathrm{t}}) \tag{2.39}$$

denotes the auxiliary unknown named rotation of the solid. Then, performing a tensor multiplication (:) by $\tau \in \mathbb{H}(\mathbf{div};\Omega)$, and applying the Green identity (1.50) through the rows of each tensor, we obtain:

$$\int_\Omega \mathscr{C}^{-1}\sigma : \tau = \int_\Omega \nabla\mathbf{u} : \tau - \int_\Omega \rho : \tau$$

$$= -\int_\Omega \mathbf{u}\cdot\mathbf{div}\,\tau + \langle\gamma_{\mathbf{n}}(\tau), \gamma_0(u)\rangle - \int_\Omega \rho : \tau,$$

where $\gamma_{\mathbf{n}} : \mathbb{H}(\mathbf{div};\Omega) \to \mathbf{H}^{-1/2}(\Gamma)$ and $\gamma_0 : \mathbf{H}^1(\Omega) \to \mathbf{H}^{1/2}(\Gamma)$ are the natural tensor and vector extensions of the respective trace operators defined in Sect. 1.3. Then, using that $\gamma_0(\mathbf{u}) = 0$ on Γ, we arrive at

$$\int_\Omega \mathscr{C}^{-1}\sigma : \tau + \int_\Omega \mathbf{u}\cdot\mathbf{div}\,\tau + \int_\Omega \rho : \tau = 0 \qquad \forall\tau \in \mathbb{H}(\mathbf{div};\Omega). \tag{2.40}$$

Note here that $\rho \in \mathbb{L}^2_{\mathrm{skew}}(\Omega)$, where

$$\mathbb{L}^2_{\mathrm{skew}}(\Omega) := \{\eta \in \mathbb{L}^2(\Omega): \quad \eta + \eta^t = 0\}.$$

In addition, it is easy to see that the symmetry of σ can be imposed weakly through the equation

$$\int_\Omega \sigma : \eta = 0 \qquad \forall \eta \in \mathbb{L}^2_{\text{skew}}(\Omega)\,. \tag{2.41}$$

Finally, the equilibrium equation is rewritten as

$$\int_\Omega \mathbf{v} \cdot \mathbf{div}\,\sigma = -\int_\Omega \mathbf{f} \cdot \mathbf{v} \qquad \forall \mathbf{v} \in \mathbf{L}^2(\Omega). \tag{2.42}$$

In this way, adding (2.41) and (2.42), and placing the resulting equation together with (2.40), we obtain the following mixed variational formulation of (2.37): find $(\sigma, (\mathbf{u}, \rho)) \in H \times Q$ such that

$$\begin{aligned} a(\sigma,\tau) + b(\tau,(\mathbf{u},\rho)) &= F(\tau) \qquad \forall \tau \in H, \\ b(\sigma,(\mathbf{v},\eta)) &= G(\mathbf{v},\eta) \qquad \forall (\mathbf{v},\eta) \in Q, \end{aligned} \tag{2.43}$$

where

$$H := \mathbb{H}(\mathbf{div};\Omega), \quad Q := \mathbf{L}^2(\Omega) \times \mathbb{L}^2_{\text{skew}}(\Omega),$$

$a : H \times H \to \mathbb{R}$, and $b : H \times Q \to \mathbb{R}$ are the bilinear forms defined by

$$a(\zeta,\tau) := \int_\Omega \mathscr{C}^{-1}\zeta : \tau = \frac{1}{2\mu}\int_\Omega \zeta : \tau - \frac{\lambda}{2\mu(n\lambda + 2\mu)}\int_\Omega \operatorname{tr}(\zeta)\operatorname{tr}(\tau), \tag{2.44}$$

for all $(\zeta,\tau) \in H \times H$, and

$$b(\tau,(\mathbf{v},\eta)) := \int_\Omega \mathbf{v} \cdot \mathbf{div}\,\tau + \int_\Omega \eta : \tau, \tag{2.45}$$

for all $(\tau,(\mathbf{v},\eta)) \in H \times Q$, and the functionals $F \in H'$ and $G \in Q'$ are given by

$$F(\tau) := 0 \quad \forall \tau \in H, \qquad G(\mathbf{v},\eta) := -\int_\Omega \mathbf{f} \cdot \mathbf{v} \quad \forall (\mathbf{v},\eta) \in Q. \tag{2.46}$$

In what follows we analyze the solvability of (2.43) by applying once again Theorem 2.3. Nevertheless, we remark in advance that this theorem will not be applied exactly to (2.43) but to an equivalent formulation. To this end, it is necessary to previously notice some properties of the bilinear forms a and b. In fact, simple algebraic computations show that

$$a(\mathbf{I},\tau) = \frac{1}{(n\lambda + 2\mu)}\int_\Omega \operatorname{tr}(\tau) \qquad \forall \tau \in H \tag{2.47}$$

and

$$a(\zeta,\tau) = \frac{1}{2\mu}\int_\Omega \zeta^{\mathrm{d}} : \tau^{\mathrm{d}} + \frac{1}{n(n\lambda + 2\mu)}\int_\Omega \operatorname{tr}(\zeta)\operatorname{tr}(\tau) \qquad \forall \zeta, \tau \in H, \tag{2.48}$$

where, given $\tau \in \mathbb{L}^2(\Omega)$, $\tau^{\mathrm{d}} := \tau - \frac{1}{n}\operatorname{tr}(\tau)\mathbf{I}$ is the corresponding deviator tensor. The convenience of writing a in the form (2.48) will become clear later on when we prove that a is elliptic in the kernel of the operator induced by b. In turn, it is easy to see that

$$b(\mathbf{I},(\mathbf{v},\eta)) = 0 \qquad \forall (\mathbf{v},\eta) \in Q. \tag{2.49}$$

We now notice that H can be decomposed as

$$H := H_0 \oplus \mathbb{R}\mathbf{I},$$

where

$$H_0 := \left\{ \tau \in \mathbb{H}(\mathbf{div};\Omega) : \quad \int_\Omega \operatorname{tr}(\tau) = 0 \right\}.$$

More specifically, for each $\tau \in H$ there exist unique

$$\tau_0 \in H_0 \quad \text{and} \quad d := \frac{1}{n|\Omega|} \int_\Omega \operatorname{tr}(\tau) \in \mathbb{R}$$

such that $\tau = \tau_0 + d\,\mathbf{I}$. As suggested by the foregoing analysis, we now consider the problem that arises from (2.43) after replacing H by H_0, that is: find $(\sigma,(\mathbf{u},\rho)) \in H_0 \times Q$ such that

$$\begin{aligned} a(\sigma,\tau) + b(\tau,(\mathbf{u},\rho)) &= F(\tau) \qquad \forall \tau \in H_0, \\ b(\sigma,(\mathbf{v},\eta)) &= G(\mathbf{v},\eta) \qquad \forall (\mathbf{v},\eta) \in Q. \end{aligned} \tag{2.50}$$

The following result connects (2.43) and (2.50).

Lemma 2.2. *Problems* (2.43) *and* (2.50) *are equivalent, that is,* $(\sigma,(\mathbf{u},\rho)) \in H \times Q$ *is a solution of* (2.43) *if and only if* $\sigma \in H_0$ *and* $(\sigma,(\mathbf{u},\rho))$ *is a solution of* (2.50).

Proof. Let $(\sigma,(\mathbf{u},\rho)) \in H \times Q$ be a solution of (2.43). Taking $\tau = \mathbf{I}$ in the first equation of (2.43), and using (2.47) and (2.49), we obtain

$$0 = a(\sigma,\mathbf{I}) + b(\mathbf{I},(\mathbf{u},\rho)) = \frac{1}{(n\lambda + 2\mu)} \int_\Omega \operatorname{tr}(\sigma),$$

which yields $\sigma \in H_0$, and hence $(\sigma,(\mathbf{u},\rho)) \in H_0 \times Q$ is a solution of (2.50).

Conversely, let $(\sigma,(\mathbf{u},\rho)) \in H_0 \times Q$ be a solution of (2.50). It is clear that $(\sigma,(\mathbf{u},\rho))$ verifies the second equation of (2.43). Now, given $\tau = \tau_0 + d\,\mathbf{I} \in H$, with $\tau_0 \in H_0$ and $d \in \mathbb{R}$, we have, using (2.49) and the first equation of (2.50), that

$$\begin{aligned} a(\sigma,\tau) + b(\tau,(\mathbf{u},\rho)) &= a(\sigma,\tau_0) + b(\tau_0,(\mathbf{u},\rho)) + d\left\{a(\sigma,\mathbf{I}) + b(\mathbf{I},(\mathbf{u},\rho))\right\} \\ &= a(\sigma,\tau_0) + b(\tau_0,(\mathbf{u},\rho)) = 0 = F(\tau), \end{aligned}$$

which shows that $(\sigma,(\mathbf{u},\rho))$ is also a solution of (2.43).

□

According to the equivalence established by the preceding lemma, we now concentrate the analysis on problem (2.50). We begin by noticing that $\mathbf{B} : H_0 \to Q$, the operator induced by $b : H_0 \times Q \to \mathbb{R}$, is given by

$$\mathbf{B}(\tau) := \left(\mathbf{div}\,\tau, \frac{1}{2}(\tau - \tau^{\mathrm{t}})\right) \qquad \forall\, \tau \in H_0, \tag{2.51}$$

from which it is clear that $\mathbf{B}$ is bounded with $\|\mathbf{B}\| \le 1$. Then, from the definition of the bilinear form a [cf. (2.44)], applying the Cauchy–Schwarz inequality, utilizing that $\|\mathrm{tr}(\tau)\|_{0,\Omega} \le \sqrt{n}\,\|\tau\|_{0,\Omega}\ \forall\, \tau \in \mathbb{L}^2(\Omega)$, and noting that $\dfrac{n\lambda}{(n\lambda + 2\mu)} \le 1$, we deduce that

$$\begin{aligned}
|a(\zeta,\tau)| &= \left|\frac{1}{2\mu}\int_\Omega \zeta : \tau - \frac{\lambda}{2\mu\,(n\lambda + 2\mu)}\int_\Omega \mathrm{tr}(\zeta)\,\mathrm{tr}(\tau)\right| \\
&\le \frac{1}{2\mu}\,\|\zeta\|_{0,\Omega}\,\|\tau\|_{0,\Omega} + \frac{1}{2\mu}\,\frac{\lambda}{(n\lambda + 2\mu)}\,\|\mathrm{tr}(\zeta)\|_{0,\Omega}\,\|\mathrm{tr}(\tau)\|_{0,\Omega} \\
&\le \frac{1}{\mu}\,\|\zeta\|_{0,\Omega}\,\|\tau\|_{0,\Omega} \le \frac{1}{\mu}\,\|\zeta\|_{\mathbf{div},\Omega}\,\|\tau\|_{\mathbf{div},\Omega} \qquad \forall\, \zeta, \tau \in H_0,
\end{aligned}$$

which proves that $\mathbf{A} : H_0 \to H_0$, the operator induced by a, is also bounded with $\|\mathbf{A}\| \le \dfrac{1}{\mu}$.

Furthermore, it is also clear from (2.51) that

$$V := N(\mathbf{B}) = \left\{\tau \in H_0 :\ \mathbf{div}\,\tau = 0 \quad \text{in} \quad \Omega, \quad \tau = \tau^{\mathrm{t}} \quad \text{in} \quad \Omega\right\},$$

and therefore, starting from (2.48), we obtain the inequality

$$a(\tau,\tau) = \frac{1}{2\mu}\,\|\tau^{\mathrm{d}}\|^2_{0,\Omega} + \frac{1}{n(n\lambda + 2\mu)}\,\|\mathrm{tr}(\tau)\|^2_{0,\Omega} \ge \frac{1}{2\mu}\,\|\tau^{\mathrm{d}}\|^2_{0,\Omega} \qquad \forall\, \tau \in V, \tag{2.52}$$

which, however, is still not enough to confirm the V-ellipticity of a. To fix this difficulty, we need the following result (cf. [16, Proposition 3.1, Chap. IV]).

Lemma 2.3. *There exists $c_1 > 0$, depending only on Ω, such that*

$$c_1\,\|\tau\|^2_{0,\Omega} \le \|\tau^{\mathrm{d}}\|^2_{0,\Omega} + \|\mathbf{div}\,\tau\|^2_{0,\Omega} \quad \forall\, \tau \in H_0.$$

Proof. We begin by recalling from [41, Corollary 2.4, Chap. I] that the divergence operator div is an isomorphism from $W^\perp$ into $L^2_0(\Omega)$, where

$$W := \left\{\mathbf{z} \in \mathbf{H}^1_0(\Omega) :\ \ \mathrm{div}\,\mathbf{z} = 0 \quad \text{in} \quad \Omega\right\}, \qquad \mathbf{H}^1_0(\Omega) = W \oplus W^\perp$$

and

$$L^2_0(\Omega) := \left\{v \in L^2(\Omega) :\ \ \int_\Omega v = 0\right\}.$$

Then, given $\tau \in H_0$, we have $\mathrm{tr}(\tau) \in L_0^2(\Omega)$ and therefore there exists a unique $\mathbf{z} \in W^\perp$ such that $\mathrm{div}\, \mathbf{z} = \mathrm{tr}(\tau)$ and

$$\|\mathbf{z}\|_{1,\Omega} \leq \tilde{C}\, \|\mathrm{tr}(\tau)\|_{0,\Omega}, \tag{2.53}$$

where $\tilde{C} > 0$ is a constant independent of $\mathbf{z}$. It follows that

$$\begin{aligned}
\|\mathrm{tr}(\tau)\|_{0,\Omega}^2 &= \int_\Omega \mathrm{tr}(\tau)\, \mathrm{div}\, \mathbf{z} = \int_\Omega \tau : \mathbf{I}\, \mathrm{tr}(\nabla \mathbf{z}) \\
&= \int_\Omega \tau : \mathrm{tr}(\nabla \mathbf{z})\, \mathbf{I} = n \int_\Omega \tau : (\nabla \mathbf{z} - (\nabla \mathbf{z})^{\mathrm{d}}) \\
&= n \int_\Omega \tau : \nabla \mathbf{z} - n \int_\Omega \tau^{\mathrm{d}} : \nabla \mathbf{z},
\end{aligned}$$

where the last expression utilizes that $\tau : \zeta^{\mathrm{d}} = \tau^{\mathrm{d}} : \zeta \quad \forall\, \tau, \zeta \in \mathbb{L}^2(\Omega)$. Hence, applying the Green identity (1.50) (cf. Lemma 1.4) in the first term, using that $\gamma_0(\mathbf{z}) = 0$ on Γ, and then employing the Cauchy–Schwarz inequality, we obtain

$$\begin{aligned}
\|\mathrm{tr}(\tau)\|_{0,\Omega}^2 &= -n \int_\Omega \mathbf{z} \cdot \mathbf{div}\, \tau - n \int_\Omega \tau^{\mathrm{d}} : \nabla \mathbf{z} \\
&\leq n \Big\{ \|\mathbf{z}\|_{0,\Omega}\, \|\mathbf{div}\, \tau\|_{0,\Omega} + \|\tau^{\mathrm{d}}\|_{0,\Omega}\, |\mathbf{z}|_{1,\Omega} \Big\} \\
&\leq n\, \|\mathbf{z}\|_{1,\Omega} \Big\{ \|\mathbf{div}\, \tau\|_{0,\Omega}^2 + \|\tau^{\mathrm{d}}\|_{0,\Omega}^2 \Big\}^{1/2},
\end{aligned}$$

which implies, together with (2.53), that

$$\|\mathrm{tr}(\tau)\|_{0,\Omega} \leq n \tilde{C} \Big\{ \|\mathbf{div}\, \tau\|_{0,\Omega}^2 + \|\tau^{\mathrm{d}}\|_{0,\Omega}^2 \Big\}^{1/2}.$$

Finally, the proof is completed by noting that

$$\|\tau\|_{0,\Omega}^2 = \|\tau^{\mathrm{d}} + \frac{1}{n} \mathrm{tr}(\tau)\, \mathbf{I}\|_{0,\Omega}^2 = \|\tau^{\mathrm{d}}\|_{0,\Omega}^2 + \frac{1}{n} \|\mathrm{tr}(\tau)\|_{0,\Omega}^2.$$

□

In this way, applying Lemma 2.3 to $\tau \in V$, we deduce from (2.52) that

$$a(\tau, \tau) \geq \frac{1}{2\mu} \|\tau^{\mathrm{d}}\|_{0,\Omega}^2 \geq \frac{c_1}{2\mu} \|\tau\|_{0,\Omega}^2 = \frac{c_1}{2\mu} \|\tau\|_{\mathbf{div},\Omega}^2,$$

which shows that a is V-elliptic with the ellipticity constant $\alpha = c_1/2\mu$.

On the other hand, given $(\mathbf{v}, \eta) \in Q$, we consider the boundary value problem

$$\mathbf{div}\, \mathbf{e}(\mathbf{z}) = \mathbf{v} - \mathbf{div}\, \eta \quad \text{in} \quad \Omega, \qquad \mathbf{z} = 0 \quad \text{on} \quad \Gamma,$$

whose primal variational formulation reads as follows: find $\mathbf{z} \in \mathbf{H}_0^1(\Omega)$ such that

$$\int_\Omega \mathbf{e}(\mathbf{z}) : \mathbf{e}(\mathbf{w}) = -\int_\Omega \mathbf{v} \cdot \mathbf{w} - \int_\Omega \eta : \nabla \mathbf{w} \qquad \forall \mathbf{w} \in \mathbf{H}_0^1(\Omega). \tag{2.54}$$

We recall here that the first Korn inequality establishes that (cf. [49, Theorem 10.1] or [14, Corollaries 9.2.22 and 9.2.25])

$$\|\mathbf{e}(\mathbf{w})\|_{0\Omega}^2 \geq \frac{1}{2} |\mathbf{w}|_{1,\Omega}^2 \qquad \forall \mathbf{w} \in \mathbf{H}_0^1(\Omega). \tag{2.55}$$

Therefore, thanks to (2.55) and the n-dimensional version of the Friedrichs–Poincaré inequality (cf. [51, Théorème 1.2-5]), a direct application of the Lax–Milgram lemma (cf. Theorem 1.1) implies that (2.54) has a unique solution $\mathbf{z} \in \mathbf{H}_0^1(\Omega)$, which, according to the corresponding continuous dependence result, satisfies

$$|\mathbf{z}|_{1,\Omega} \leq \tilde{C} \left\{ \|\mathbf{v}\|_{0,\Omega} + \|\eta\|_{0,\Omega} \right\}. \tag{2.56}$$

Then, defining $\hat{\tau} := \mathbf{e}(\mathbf{z}) + \eta$ in Ω, we clearly have $\hat{\tau} \in \mathbb{L}^2(\Omega)$ and $\mathbf{div}\,\hat{\tau} = \mathbf{v}$ in Ω, and thus $\hat{\tau} \in \mathbb{H}(\mathbf{div};\Omega)$. In this way, denoting by $\tilde{\tau}$ the H_0-component of $\hat{\tau}$, we find that

$$\mathbf{B}(\tilde{\tau}) := \left(\mathbf{div}\,\tilde{\tau}, \frac{1}{2}(\tilde{\tau} - \tilde{\tau}^{\mathtt{t}}) \right) = (\mathbf{v}, \eta),$$

which proves that $\mathbf{B}$ is surjective. Notice here, using the identity $\mathbf{e}(\mathbf{z}) : \eta = 0$ and the inequality (2.56), that

$$\begin{aligned} \|\tilde{\tau}\|_{\mathbf{div},\Omega}^2 \leq \|\hat{\tau}\|_{\mathbf{div},\Omega}^2 &= \|\mathbf{e}(\mathbf{z})\|_{0,\Omega}^2 + \|\eta\|_{0,\Omega}^2 + \|\mathbf{v}\|_{0,\Omega}^2 \\ &\leq (1 + 2\tilde{C}^2)\, \|(\mathbf{v}, \eta)\|_Q^2. \end{aligned} \tag{2.57}$$

Finally, it follows straightforwardly from (2.46) that G is bounded with

$$\|G\| \leq \|\mathbf{f}\|_{0,\Omega}.$$

Therefore, Theorem 2.3 guarantees that problem (2.50) has a unique solution $(\sigma, (\mathbf{u}, \rho)) \in H_0 \times Q$, and there holds

$$\|(\sigma, (\mathbf{u}, \rho))\|_{H \times Q} \leq C \|\mathbf{f}\|_{0,\Omega},$$

with $C > 0$ depending on the constant β for the continuous inf-sup condition for b, $\|\mathbf{A}\| \leq 1/\mu$, and the ellipticity constant $\alpha = c_1/2\mu$. We show next that estimate (2.57) allows us to obtain a closer look at β. In fact, given $(\mathbf{v}, \eta) \in Q$, we have

$$\sup_{\substack{\tau \in H_0 \\ \tau \neq 0}} \frac{b(\tau, (\mathbf{v}, \eta))}{\|\tau\|_H} \geq \frac{b(\tilde{\tau}, (\mathbf{v}, \eta))}{\|\tilde{\tau}\|_H} = \frac{\|(\mathbf{v}, \eta)\|_Q^2}{\|\tilde{\tau}\|_H} \geq \frac{1}{(1 + 2\tilde{C}^2)^{1/2}} \|(\mathbf{v}, \eta)\|_Q,$$

which implies that we can set $\beta := \dfrac{1}{(1+2\tilde{C}^2)^{1/2}}$.

Before analyzing the problem with mixed boundary conditions, we find it interesting to discuss the situation of nonhomogeneous Dirichlet boundary conditions. In this case, given $\mathbf{g} \in \mathbf{H}^{1/2}(\Gamma)$, the functional F in (2.43) and (2.50) becomes

$$F(\tau) := \langle \gamma_{\mathbf{n}}(\tau), \mathbf{g} \rangle \qquad \forall \tau \in H, \tag{2.58}$$

and the equivalence given by Lemma 2.2 needs to be modified as follows.

Lemma 2.4. *Problems* (2.43) *and* (2.50)*, with F given by* (2.58)*, are equivalent in the following sense:*

(i) *Let $(\sigma, (\mathbf{u}, \rho)) \in H \times Q$ be a solution of* (2.43)*, and let $\sigma := \sigma_0 + c\mathbf{I}$, with $\sigma_0 \in H_0$ and $c \in \mathbb{R}$. Then $(\sigma_0, (\mathbf{u}, \rho)) \in H_0 \times Q$ is a solution of* (2.50)*.*
(ii) *Let $(\sigma_0, (\mathbf{u}, \rho)) \in H_0 \times Q$ be a solution of* (2.50)*, and define $\sigma := \sigma_0 + c\mathbf{I}$, with $c := \dfrac{(n\lambda + 2\mu)}{n|\Omega|} \displaystyle\int_\Gamma \mathbf{g} \cdot \mathbf{n}$. Then $(\sigma, (\mathbf{u}, \rho)) \in H \times Q$ is a solution of* (2.43)*.*

Proof. The proof is similar to that of Lemma 2.2. We omit further details and leave it as an exercise for the reader. □

2.4.3.2 Mixed Boundary Conditions

We now return to the general case given by (2.35), that is,

$$\begin{aligned} \sigma = \mathscr{C}\mathbf{e}(\mathbf{u}) \quad \text{in} \quad \Omega, \qquad \mathbf{div}\,\sigma = -\mathbf{f} \quad \text{in} \quad \Omega, \\ \mathbf{u} = 0 \quad \text{on} \quad \Gamma_D, \qquad \sigma\mathbf{n} = \mathbf{g} \quad \text{on} \quad \Gamma_N, \end{aligned} \tag{2.59}$$

where $\mathbf{f} \in \mathbf{L}^2(\Omega)$ and $\mathbf{g} \in \mathbf{H}_{00}^{-1/2}(\Gamma_N)$. Adopting the same procedure from the previous cases, that is, inverting Hooke's law as in (2.38), defining the rotation ρ [cf. (2.39)] as an additional unknown, and applying the Green identity (1.50), we first obtain for each $\tau \in \mathbb{H}(\mathbf{div};\Omega)$

$$\int_\Omega \mathscr{C}^{-1}\sigma : \tau = -\int_\Omega \mathbf{u} \cdot \mathbf{div}\,\tau + \langle \gamma_{\mathbf{n}}(\tau), \gamma_0(\mathbf{u}) \rangle - \int_\Omega \rho : \tau. \tag{2.60}$$

Then, defining the auxiliary unknown $\xi := -\gamma_0(\mathbf{u}) \in \mathbf{H}_{00}^{1/2}(\Gamma_N)$, which is supported by the homogeneous condition $\gamma_0(\mathbf{u}) = 0$ on Γ_D, we deduce from (2.60) that

$$\int_\Omega \mathscr{C}^{-1}\sigma : \tau + \int_\Omega \mathbf{u} \cdot \mathbf{div}\,\tau + \int_\Omega \rho : \tau + \langle \gamma_{\mathbf{n}}(\tau)|_{\Gamma_N}, \xi \rangle_{\Gamma_N} = 0 \quad \forall \tau \in \mathbb{H}(\mathbf{div};\Omega),$$

where $\langle \cdot, \cdot \rangle_{\Gamma_N}$ stands for the duality between $\mathbf{H}_{00}^{-1/2}(\Gamma_N)$ and $\mathbf{H}_{00}^{1/2}(\Gamma_N)$. For the remaining equations we proceed as in Sect. 2.4.2 [cf. (2.29) and (2.30)] and finally arrive at the following mixed variational formulation of (2.59): find $(\sigma, (\mathbf{u}, \rho, \xi)) \in H \times Q$ such that

$$\begin{aligned} a(\sigma,\tau) + b(\tau,(\mathbf{u},\rho,\xi)) &= F(\tau) \quad \forall \tau \in H, \\ b(\sigma,(\mathbf{v},\eta,\chi)) &= G(\mathbf{v},\eta,\chi) \quad \forall (\mathbf{v},\eta,\chi) \in Q, \end{aligned} \tag{2.61}$$

where

$$H := \mathbb{H}(\mathbf{div};\Omega), \quad Q := \mathbf{L}^2(\Omega) \times \mathbb{L}^2_{\text{skew}}(\Omega) \times \mathbf{H}^{1/2}_{00}(\Gamma_N),$$

$a : H \times H \to \mathbb{R}$ is the bilinear form given by (2.44), $b : H \times Q \to \mathbb{R}$ is defined by

$$b(\tau,(\mathbf{v},\eta,\chi)) := \int_\Omega \mathbf{v} \cdot \mathbf{div}\,\tau + \int_\Omega \eta : \tau + \langle \gamma_{\mathbf{n}}(\tau)|_{\Gamma_N}, \chi \rangle_{\Gamma_N} \tag{2.62}$$

for all $(\tau,(\mathbf{v},\eta,\chi)) \in H \times Q$, $F \in H'$ is the null functional [cf. (2.46)], and $G \in Q'$ is given by

$$G(\mathbf{v},\eta,\chi) := -\int_\Omega \mathbf{f} \cdot \mathbf{v} + \langle \mathbf{g}, \chi \rangle_{\Gamma_N} \qquad \forall (\mathbf{v},\eta,\chi)) \in Q. \tag{2.63}$$

We remark that, unlike the statement of Lemma 2.2, it is not possible now to equivalently reformulate (2.61) with H_0 instead of H. Indeed, from Hooke's law (2.36) and the first equation of (2.59) we easily obtain

$$\text{tr}(\sigma) = (n\lambda + 2\mu)\,\text{tr}\big(\mathbf{e}(\mathbf{u})\big) = (n\lambda + 2\mu)\,\mathbf{div}\,\mathbf{u},$$

whence, using that $\mathbf{u} = 0$ on Γ_D, we find that

$$\int_\Omega \text{tr}(\sigma) = (n\lambda + 2\mu) \int_\Omega \text{div}\,\mathbf{u} = (n\lambda + 2\mu) \int_{\Gamma_N} \mathbf{u} \cdot \mathbf{n}.$$

Equivalently, taking $\tau = \mathbf{I}$ in the first equation of (2.61), and using (2.47), we have

$$\frac{1}{(n\lambda + 2\mu)} \int_\Omega \text{tr}(\sigma) + \langle \mathbf{n}, \xi \rangle_{\Gamma_N} = 0,$$

that is, recalling that $\xi = -\gamma_0(\mathbf{u})$ on Γ_N,

$$\int_\Omega \text{tr}(\sigma) = (n\lambda + 2\mu)\,\langle \mathbf{n}, \gamma_0(\mathbf{u}) \rangle_{\Gamma_N} = (n\lambda + 2\mu) \int_{\Gamma_N} \mathbf{u} \cdot \mathbf{n}.$$

In other words, the presence of the Neumann boundary condition on Γ_N does not allow us to conclude that $\sigma \in H_0$, and hence the inequality provided by Lemma 2.3 cannot be applied in this case. To circumvent this difficulty, we now prove the following result.

Lemma 2.5. *There exists $c_2 > 0$, depending on Γ_N and Ω, such that*

$$c_2 \|\tau\|^2_{\mathbf{div},\Omega} \le \|\tau_0\|^2_{\mathbf{div},\Omega} \qquad \forall \tau := \tau_0 + d\,\mathbf{I} \in \mathbb{H}_{\Gamma_N}(\mathbf{div};\Omega),$$

with $\tau_0 \in H_0$ and $d \in \mathbb{R}$, where

$$\mathbb{H}_{\Gamma_N}(\mathbf{div};\Omega) := \big\{ \tau \in \mathbb{H}(\mathbf{div};\Omega) : \quad \gamma_{\mathbf{n}}(\tau)|_{\Gamma_N} = 0 \big\}.$$

Proof. Let $\tau := \tau_0 + d\,\mathbf{I} \in \mathbb{H}_{\Gamma_N}(\mathbf{div};\Omega)$, with $\tau_0 \in H_0$ and $d \in \mathbb{R}$. It follows that

$$0 = \gamma_{\mathbf{n}}(\tau)|_{\Gamma_N} = \gamma_{\mathbf{n}}(\tau_0)|_{\Gamma_N} + d\,\gamma_{\mathbf{n}}(\mathbf{I})|_{\Gamma_N} = \gamma_{\mathbf{n}}(\tau_0)|_{\Gamma_N} + d\,\mathbf{n},$$

and hence, using that

$$\|\psi|_{\Gamma_N}\|_{-1/2,00,\Gamma_N} \leq \|\psi\|_{-1/2,\Gamma} \qquad \forall\, \psi \in \mathbf{H}^{-1/2}(\Gamma),$$

which is the vector version of (2.27), we deduce that

$$|d|\,\|\mathbf{n}\|_{-1/2,00,\Gamma_N} = \|\gamma_{\mathbf{n}}(\tau_0)|_{\Gamma_N}\|_{-1/2,00,\Gamma_N} \leq \|\gamma_{\mathbf{n}}(\tau_0)\|_{-1/2,\Gamma} \leq \|\tau_0\|_{\mathbf{div},\Omega},$$

that is,

$$|d| \leq \frac{1}{\|\mathbf{n}\|_{-1/2,00,\Gamma_N}}\,\|\tau_0\|_{\mathbf{div},\Omega}.$$

The preceding inequality, together with the fact that

$$\|\tau\|^2_{\mathbf{div},\Omega} = \|\tau_0\|^2_{\mathbf{div},\Omega} + n\,d^2|\Omega|,$$

implies that

$$\|\tau\|^2_{\mathbf{div},\Omega} \leq \left\{1 + \frac{n|\Omega|}{\|\mathbf{n}\|^2_{-1/2,00,\Gamma_N}}\right\}\|\tau_0\|^2_{\mathbf{div},\Omega},$$

which completes the proof. □

We now observe that $\mathbf{B} : H \to Q$, the operator induced by the present bilinear form $b : H \times Q \to \mathbb{R}$ [cf. (2.62)], is given by

$$\mathbf{B}(\tau) := \left(\mathbf{div}\,\tau,\ \frac{1}{2}(\tau - \tau^{\mathtt{t}}),\ \mathscr{R}_{00}\,\gamma_{\mathbf{n}}(\tau)|_{\Gamma_N}\right) \quad \forall\, \tau \in H,$$

where $\mathscr{R}_{00} : \mathbf{H}^{-1/2}_{00}(\Gamma_N) \to \mathbf{H}^{1/2}_{00}(\Gamma_N)$ is the corresponding Riesz mapping. Like the analysis in Sect. 2.4.2, it is shown here that $\mathbf{B}$ is bounded with $\|\mathbf{B}\| \leq 2$. In addition, it is easy to see that

$$V := N(\mathbf{B}) = \left\{\tau \in H :\ \mathbf{div}\,\tau = 0 \ \text{ in } \ \Omega,\quad \tau = \tau^{\mathtt{t}} \ \text{ in } \ \Omega,\quad \gamma_{\mathbf{n}}(\tau)|_{\Gamma_N} = 0\right\},$$

and hence, given $\tau \in V$, we apply (2.52) and Lemmas 2.3 and 2.5 to deduce that

$$\begin{aligned} a(\tau,\tau) &\geq \frac{1}{2\mu}\,\|\tau^{\mathtt{d}}\|^2_{0,\Omega} = \frac{1}{2\mu}\,\|\tau_0^{\mathtt{d}}\|^2_{0,\Omega} \geq \frac{c_1}{2\mu}\,\|\tau_0\|^2_{0,\Omega} \\ &= \frac{c_1}{2\mu}\,\|\tau_0\|^2_{\mathbf{div},\Omega} \geq \frac{c_1\,c_2}{2\mu}\,\|\tau\|^2_{\mathbf{div},\Omega}, \end{aligned}$$

which proves that a is V-elliptic with constant $\alpha = c_1\,c_2/2\mu$.

Furthermore, given $(\mathbf{v},\eta,\chi) \in Q$, we consider the boundary value problem

$$\mathbf{div}\,\mathbf{e}(\mathbf{z}) = \mathbf{v} - \mathbf{div}\,\eta \quad \text{in} \quad \Omega, \quad \mathbf{z} = 0 \quad \text{on} \quad \Gamma_D,$$
$$(\mathbf{e}(\mathbf{z}) + \eta)\mathbf{n} = \mathscr{R}_{00}^{-1}(\chi) \quad \text{on} \quad \Gamma_N,$$

whose primal formulation reads as follows: find $\mathbf{z} \in \mathbf{H}^1_{\Gamma_D}(\Omega)$ such that

$$\int_\Omega \mathbf{e}(\mathbf{z}) : \mathbf{e}(\mathbf{w}) = -\int_\Omega \mathbf{v}\cdot\mathbf{w} - \int_\Omega \eta : \nabla\mathbf{w} + \langle \mathscr{R}_{00}^{-1}(\chi), \gamma_0(\mathbf{w})\rangle_{\Gamma_N} \quad \forall\,\mathbf{w} \in \mathbf{H}^1_{\Gamma_D}(\Omega), \tag{2.64}$$

where $\mathbf{H}^1_{\Gamma_D}(\Omega) := \{\mathbf{w} \in \mathbf{H}^1(\Omega) : \quad \gamma_0(\mathbf{w}) = 0 \quad \text{on} \quad \Gamma_D\}$. In this case, the Korn inequality (cf. [14]) establishes the existence of $c > 0$ such that

$$\|\mathbf{e}(\mathbf{w})\|^2_{0,\Omega} \geq c\,|\mathbf{w}|^2_{1,\Omega} \qquad \forall\,\mathbf{w} \in \mathbf{H}^1_{\Gamma_D}(\Omega),$$

and hence, the Lax–Milgram lemma (cf. Theorem 1.1) implies that (2.64) has a unique solution $\mathbf{z} \in \mathbf{H}^1_{\Gamma_D}(\Omega)$ satisfying

$$|\mathbf{z}|_{1,\Omega} \leq \hat{C}\left\{\|\mathbf{v}\|_{0,\Omega} + \|\eta\|_{0,\Omega} + \|\chi\|_{1/2,00,\Gamma_N}\right\}, \tag{2.65}$$

where $\hat{C} > 0$ depends on c and the constant that arises from the equivalence between $\|\cdot\|_{1,\Omega}$ and $|\cdot|_{1,\Omega}$ in $\mathbf{H}^1_{\Gamma_D}(\Omega)$, which is also a consequence of the generalized Poincaré inequality (cf. [46, Theorem 5.11.2]). In this way, defining $\hat{\tau} := \mathbf{e}(\mathbf{z}) + \eta$ in Ω, we clearly have $\hat{\tau} \in \mathbb{L}^2(\Omega)$ and $\mathbf{div}\,\hat{\tau} = \mathbf{v}$ in Ω, which implies $\hat{\tau} \in \mathbb{H}(\mathbf{div};\Omega)$, and then $\gamma_{\mathbf{n}}(\hat{\tau})|_{\Gamma_N} = \mathscr{R}_{00}^{-1}(\chi)$. It follows that $\mathbf{B}(\hat{\tau}) = (\mathbf{v},\eta,\chi)$, thus showing that $\mathbf{B}$ is surjective. Furthermore, using the fact that $\mathbf{e}(\mathbf{z}) : \eta = 0$ and employing inequality (2.65), we obtain

$$\|\hat{\tau}\|^2_{\mathbf{div},\Omega} = \|\mathbf{e}(\mathbf{z})\|^2_{0,\Omega} + \|\eta\|^2_{0,\Omega} + \|\mathbf{v}\|^2_{0,\Omega} \leq (1 + 3\hat{C}^2)\,\|(\mathbf{v},\eta,\chi)\|^2_Q. \tag{2.66}$$

On the other hand, it is clear from (2.63) that G is bounded with

$$\|G\| \leq \|\mathbf{f}\|_{0,\Omega} + \|\mathbf{g}\|_{-1/2,00,\Gamma_N}.$$

Consequently, a direct application of Theorem 2.3 guarantees that (2.61) has a unique solution $(\sigma,(\mathbf{u},\rho,\xi)) \in H \times Q$, which satisfies

$$\|(\sigma,(\mathbf{u},\rho,\xi))\|_{H\times Q} \leq C\left\{\|\mathbf{f}\|_{0,\Omega} + \|\mathbf{g}\|_{-1/2,00,\Gamma_N}\right\},$$

with $C > 0$ depending on the constant β for the continuous inf-sup condition of b, $\|\mathbf{A}\| \leq 1/\mu$, and the ellipticity constant $\alpha = c_1 c_2/2\mu$. Finally, to estimate β, we denote by $\langle\cdot,\cdot\rangle_Q$ the inner product of Q, and observe, thanks to (2.66), that

$$\sup_{\substack{\tau \in H \\ \tau \neq 0}} \frac{b(\tau,(\mathbf{v},\eta,\chi))}{\|\tau\|_H} \geq \frac{b(\hat{\tau},(\mathbf{v},\eta,\chi))}{\|\hat{\tau}\|_{\mathbf{div},\Omega}} = \frac{\langle \mathbf{B}(\hat{\tau}),(\mathbf{v},\eta,\chi)\rangle_Q}{\|\hat{\tau}\|_{\mathbf{div},\Omega}}$$

$$= \frac{\|(\mathbf{v},\eta,\chi)\|_Q^2}{\|\hat{\tau}\|_{\mathbf{div},\Omega}} \geq \frac{1}{(1+3\hat{C}^2)^{1/2}} \|(\mathbf{v},\eta,\chi)\|_Q ,$$

which suggests setting $\beta := \dfrac{1}{(1+3\hat{C}^2)^{1/2}}$.

We end this section by mentioning that a complete analysis of the linear elasticity problem with pure Neumann boundary conditions can be found in [35].

2.4.4 Primal-Mixed Formulation of Poisson Problem

The present set of application examples is completed by the Poisson problem analyzed in Sect. 2.4.1, but utilizing now what we call the primal-mixed formulation. Recall that the geometry is given by a bounded domain Ω of $\mathbb{R}^n$, $n \geq 2$, with Lipschitz-continuous boundary Γ. Then, given $f \in L^2(\Omega)$ and $g \in H^{1/2}(\Gamma)$, we are interested in the following boundary value problem:

$$-\Delta u = f \quad \text{in} \quad \Omega, \qquad u = g \quad \text{on} \quad \Gamma. \tag{2.67}$$

Multiplying the partial differential equation by $v \in H^1(\Omega)$ and applying the improved Green identity (1.52) (cf. Theorem 1.8), we obtain

$$\int_\Omega f v = -\int_\Omega \Delta u = \int_\Omega \nabla u \cdot \nabla v - \langle \gamma_1(u), \gamma_0(v)\rangle ,$$

where $\gamma_1 : H^1_\Delta(\Omega) \to H^{-1/2}(\Gamma)$ is the linear and bounded operator given by $\gamma_{\mathbf{n}} \circ \nabla$, and $\langle \cdot,\cdot \rangle$ denotes the duality between $H^{-1/2}(\Gamma)$ and $H^{1/2}(\Gamma)$. Then, introducing the auxiliary unknown $\xi := -\gamma_1(u) \in H^{-1/2}(\Gamma)$, we can write

$$\int_\Omega \nabla u \cdot \nabla v + \langle \xi, \gamma_0(v)\rangle = \int_\Omega f v \qquad \forall v \in H^1(\Omega). \tag{2.68}$$

Subsequently, the nonhomogeneous Dirichlet boundary condition $\gamma_0(u) = g$ is weakly imposed as

$$\langle \lambda, \gamma_0(u)\rangle = \langle \lambda, g\rangle \qquad \forall \lambda \in H^{-1/2}(\Gamma). \tag{2.69}$$

In this way, placing together (2.68) and (2.69), we arrive at the primal-mixed variational formulation of (2.67): find $(u,\xi) \in H \times Q$ such that

$$\begin{aligned} a(u,v) + b(v,\xi) &= F(v) \quad \forall v \in H, \\ b(u,\lambda) \qquad\qquad &= G(\lambda) \quad \forall \lambda \in Q, \end{aligned} \tag{2.70}$$

where $H := H^1(\Omega)$, $Q := H^{-1/2}(\Gamma)$, a and b are the bilinear forms defined by

$$a(u,v) := \int_\Omega \nabla u \cdot \nabla v \qquad \forall (u,v) \in H \times H,$$

$$b(v,\lambda) := \langle \lambda, \gamma_0(v) \rangle \qquad \forall (v,\lambda) \in H \times Q,$$

and the functionals $F \in H'$ and $G \in Q'$ are given by

$$F(v) := \int_\Omega f v \quad \forall v \in H, \qquad G(\lambda) := \langle \lambda, g \rangle \quad \forall \lambda \in Q.$$

It is clear that a and b are bounded since, applying the Cauchy–Schwarz inequality, we obtain

$$|a(u,v)| \le |u|_{1,\Omega} \, |v|_{1,\Omega} \le \|u\|_{1,\Omega} \, \|v\|_{1,\Omega},$$

and employing also the trace inequality we have

$$|b(v,\lambda)| \le \|\lambda\|_{-1/2,\Gamma} \, \|\gamma_0(v)\|_{1/2,\Gamma} \le \|v\|_{1,\Omega} \, \|\lambda\|_{-1/2,\Gamma},$$

which shows that $\|\mathbf{A}\| \le 1$ and $\|\mathbf{B}\| \le 1$, where $\mathbf{A} : H \to H$ and $\mathbf{B} : H \to Q$ are the operators induced by a and b, respectively. Moreover, if $\mathscr{R} : H^{-1/2}(\Gamma) \to H^{1/2}(\Gamma)$ denotes the corresponding Riesz mapping, then we have

$$b(v,\lambda) = \langle \lambda, \gamma_0(v) \rangle = \langle \mathscr{R}(\lambda), \gamma_0(v) \rangle_{1/2,\Gamma} = \langle \mathscr{R}^* \gamma_0(v), \lambda \rangle_{-1/2,\Gamma},$$

where $\langle \cdot, \cdot \rangle_{r,\Gamma}$ is the inner product of $H^r(\Gamma)$, $r \in \{-1/2, 1/2\}$, which shows that the operator $\mathbf{B}$ reduces to

$$\mathbf{B}(v) = \mathscr{R}^* \gamma_0(v) \qquad \forall v \in H.$$

Thus, since the adjoint $\mathscr{R}^* : H^{1/2}(\Gamma) \to H^{-1/2}(\Gamma)$ is certainly bijective, it follows that

$$V := \mathbf{N}(\mathbf{B}) = \Big\{ v \in H : \mathbf{B}(v) = 0 \Big\} = \Big\{ v \in H^1(\Omega) : \gamma_0(v) = 0 \Big\} = H_0^1(\Omega),$$

and hence, thanks to the Friedrichs–Poincaré inequality, there exists $\alpha > 0$ such that

$$a(v,v) = |v|_{1,\Omega}^2 \ge \alpha \, \|v\|_{1,\Omega}^2 \qquad \forall v \in V,$$

which proves the V-ellipticity of a.

On the other hand, for the surjectivity of $\mathbf{B}$ it suffices to see that this operator is given by the composition of the operators $\mathscr{R}^*$ (which is bijective) and γ_0 (which is surjective). For example, given $\lambda \in H^{-1/2}(\Gamma)$, we have that $z := \tilde{\gamma}_0^{-1}(\mathscr{R}^*)^{-1}(\lambda) \in H_0^1(\Omega)^\perp$ satisfies $\mathbf{B}(z) = \lambda$, confirming the preceding assertion. Finally, utilizing the Cauchy–Schwarz inequality and the duality between $H^{-1/2}(\Gamma)$ and $H^{1/2}(\Gamma)$, it follows easily that F and G are bounded with $\|F\| \le \|f\|_{0,\Omega}$ and $\|G\| \le \|g\|_{1/2,\Gamma}$.

Consequently, applying once again Theorem 2.3, we deduce that there exists a unique $(u,\xi) \in H \times Q$ solution of (2.70) that satisfies

$$\|(u,\xi)\|_{H\times Q} \le C\Big\{\|f\|_{0,\Omega} + \|g\|_{1/2,\Gamma}\Big\},$$

with $C > 0$ depending on the constant β for the continuous inf-sup condition of b, $\|\mathbf{A}\| \le 1$, and the ellipticity constant α. Then, to obtain an explicit value for β, we proceed as in all the previous examples. Indeed, given $\lambda \in Q := H^{-1/2}(\Gamma)$, we have, making use of (1.36) (cf. Lemma 1.3), that

$$\sup_{\substack{v\in H\\ v\neq 0}} \frac{b(v,\lambda)}{\|v\|_H} = \sup_{\substack{v\in H^1(\Omega)\\ v\neq 0}} \frac{\langle \lambda, \gamma_0(v)\rangle}{\|v\|_{1,\Omega}} \ge \frac{\langle \lambda, \gamma_0\,\tilde{\gamma}_0^{-1}(\mathscr{R}\lambda)\rangle}{\|\tilde{\gamma}_0^{-1}(\mathscr{R}(\lambda))\|_{1,\Omega}}$$
$$= \frac{\langle \lambda, \mathscr{R}(\lambda)\rangle}{\|\tilde{\gamma}_0^{-1}(\mathscr{R}(\lambda))\|_{1,\Omega}} = \frac{\|\mathscr{R}(\lambda)\|_{1/2,\Gamma}^2}{\|\mathscr{R}(\lambda)\|_{1/2,\Gamma}} = \|\mathscr{R}(\lambda)\|_{1/2,\Gamma} = \|\lambda\|_{-1/2,\Gamma},$$

which yields $\beta = 1$.

For further applications in continuum mechanics of the Babuška–Brezzi theory and related abstract developments, we refer readers to the classic books [16, 41] and to the recent updated version of [16] given by [13], which, among several new features, provides interesting new results on electromagnetism problems.

2.5 Galerkin Scheme

Let $\{H_h\}_{h>0}$ and $\{Q_h\}_{h>0}$ be sequences of finite-dimensional subspaces of H and Q, respectively. Then, given $F \in H'$ and $G \in Q'$, the Galerkin scheme of (2.1) reads as follows: find $(\sigma_h, u_h) \in H_h \times Q_h$ such that

$$\begin{aligned} a(\sigma_h,\tau_h) + b(\tau_h,u_h) &= F(\tau_h) \qquad \forall\, \tau_h \in H_h,\\ b(\sigma_h,v_h) &= G(v_h) \qquad \forall\, v_h \in Q_h. \end{aligned} \tag{2.71}$$

For the analysis of (2.71) we basically follow the same approach of Sect. 2.1. In fact, let $\mathbf{A}_h : H_h \to H_h$ and $\mathbf{B}_h : H_h \to Q_h$ be the linear and bounded operators induced by a and b on $H_h \times H_h$ and $H_h \times Q_h$, respectively, that is,

$$\mathbf{A}_h := \mathscr{R}_{H_h} \circ \mathscr{A}_h \quad \text{and} \quad \mathbf{B}_h := \mathscr{R}_{Q_h} \circ \mathscr{B}_h,$$

where $\mathscr{R}_{H_h} : H_h' \to H_h$ and $\mathscr{R}_{Q_h} : Q_h' \to Q_h$ are the corresponding Riesz mappings, and the operators $\mathscr{A}_h : H_h \to H_h'$ and $\mathscr{B}_h : Q_h \to Q_h'$ are defined by

$$\mathscr{A}_h(\sigma_h)(\tau_h) := a(\sigma_h,\tau_h) \qquad \forall\, \sigma_h \in H_h, \quad \forall\, \tau_h \in H_h$$

and

$$\mathscr{B}_h(\tau_h)(v_h) := b(\tau_h, v_h) \qquad \forall \tau_h \in H_h, \quad \forall v_h \in Q_h.$$

The following theorem establishes sufficient conditions for (2.71) to be well-posed.

Theorem 2.4. *Let* $V_h := N(\mathbf{B}_h) = \{ \tau_h \in H_h : \quad b(\tau_h, v_h) = 0 \quad \forall v_h \in Q_h \}$, *and let* $\Pi_h : H_h \to V_h$ *be the orthogonal projection operator. Assume that:*

(i) $\Pi_h \mathbf{A}_h : V_h \to V_h$ *is injective;*
(ii) *There exists* $\beta_h > 0$ *such that*

$$\sup_{\substack{\tau_h \in H_h \\ \tau_h \neq 0}} \frac{b(\tau_h, v_h)}{\|\tau_h\|_H} \geq \beta_h \|v_h\|_Q \qquad \forall v_h \in Q_h. \tag{2.72}$$

Then for each pair $(F,G) \in H' \times Q'$ *there exists a unique* $(\sigma_h, u_h) \in H_h \times Q_h$ *solution of* (2.71). *Moreover, there exists a constant* $C_h > 0$, *which depends on* $\|\mathbf{A}_h\|$, $\|(\Pi_h \mathbf{A}_h)^{-1}\|$ *and* β_h, *such that*

$$\|(\sigma_h, u_h)\|_{H \times Q} \leq C_h \left\{ \|F_h\|_{H_h'} + \|G_h\|_{Q_h'} \right\}, \tag{2.73}$$

where $F_h := F|_{H_h}$ *and* $G_h := G|_{Q_h}$.

Proof. Note first that, since V_h is of finite dimension, $\Pi_h \mathbf{A}_h : V_h \to V_h$ is injective if and only if it is surjective, and therefore hypothesis (i) is equivalent to requiring that this operator be bijective. Consequently, the rest of the proof is just a simple application of Theorem 2.1 to the present discrete setting.

□

Analogously to the proof of Theorem 2.2, one can prove here that (i) and (ii) from Theorem 2.4 are also **necessary** conditions. In addition, it is easy to see that (i) is equivalent to each of the following inf-sup conditions for a:

(i-1) There exists $\alpha_h > 0$ such that

$$\sup_{\substack{\tau_h \in V_h \\ \tau_h \neq 0}} \frac{a(\sigma_h, \tau_h)}{\|\tau_h\|_H} \geq \alpha_h \|\sigma_h\|_H \qquad \forall \sigma_h \in V_h; \tag{2.74}$$

(i-1)' There exists $\alpha_h > 0$ such that

$$\sup_{\substack{\tau_h \in V_h \\ \tau_h \neq 0}} \frac{a(\tau_h, \sigma_h)}{\|\tau_h\|_H} \geq \alpha_h \|\sigma_h\|_H \qquad \forall \sigma_h \in V_h. \tag{2.75}$$

Since (2.74), (2.75), and (2.72) hold in finite-dimensional spaces, they are usually called DISCRETE INF-SUP CONDITIONS for a and b, respectively.

Then, as was observed for the V-ellipticity of a, one can show that a sufficient (but not necessary) condition for (i) is the V_h-ellipticity of the bilinear form a, which means assuming the existence of $\alpha_h > 0$ such that

$$a(\tau_h, \tau_h) \geq \alpha_h \|\tau_h\|_H^2 \qquad \forall \tau_h \in V_h. \tag{2.76}$$

On the other hand, it is important to remark that one usually refers to (2.73) as the STABILITY OF THE GALERKIN SCHEME (2.71). Moreover, it is pretty straightforward to see that this inequality implies the boundedness of the GALERKIN PROJECTOR

$$G_h : H \times Q \to H_h \times Q_h,$$

which, given $(\zeta, w) \in H \times Q$, is defined by $G_h(\zeta, w) := (\zeta_h, w_h)$, where $(\zeta_h, w_h) \in H_h \times Q_h$ is the unique solution of the Galerkin problem

$$\begin{aligned} a(\zeta_h, \tau_h) + b(\tau_h, w_h) &= a(\zeta, \tau_h) + b(\tau_h, w) \qquad \forall \tau_h \in H_h, \\ b(\zeta_h, v_h) \qquad\qquad &= b(\zeta, v_h) \qquad \forall v_h \in Q_h. \end{aligned} \tag{2.77}$$

Indeed, it follows from (2.73) and (2.77) that $\|G_h\|$ depends on $\|\mathbf{A}_h\|$, $\|(\Pi_h \mathbf{A}_h)^{-1}\|$, β_h, $\|\mathbf{A}\|$, and $\|\mathbf{B}\|$. Note also that

$$G_h(\zeta_h, w_h) = (\zeta_h, w_h) \quad \forall (\zeta_h, w_h) \in H_h \times Q_h \tag{2.78}$$

and

$$G_h(\sigma, u) = (\sigma_h, u_h), \tag{2.79}$$

where $(\sigma, u) \in H \times Q$ and $(\sigma_h, u_h) \in H_h \times Q_h$ are the unique solutions of (2.1) and (2.71), respectively.

The boundedness of G_h allows us to easily derive the a priori estimate for the error $\|(\sigma, u) - (\sigma_h, u_h)\|_{H \times Q}$, which is known as the Cea estimate.

Theorem 2.5. *Under the assumptions of Theorems* 2.1 *and* 2.4, *there holds*

$$\|(\sigma, u) - (\sigma_h, u_h)\|_{H \times Q} \leq \|G_h\| \inf_{(\zeta_h, w_h) \in H_h \times Q_h} \|(\sigma, u) - (\zeta_h, w_h)\|_{H \times Q}.$$

Proof. It suffices to see, utilizing (2.78) and (2.79), that

$$(\sigma, u) - (\sigma_h, u_h) = (I - G_h)\Big((\sigma, u) - (\zeta_h, w_h)\Big) \qquad \forall (\zeta_h, w_h) \in H_h \times Q_h,$$

where I is the identity operator, and then apply a recent result (cf. [55]) establishing, since G_h is a projector, that $\|I - G_h\| = \|G_h\|$.

□

Certainly, to confirm the convergence of the Galerkin scheme, that is,

$$\lim_{h \to 0} \|(\sigma, u) - (\sigma_h, u_h)\|_{H \times Q} = 0,$$

$\|G_h\|$ must be independent of h, which means requiring that all the constants involved, including the norms of the operators and the discrete inf-sup conditions, be independent of the subspace $H_h \times Q_h$. Actually, this necessity of the independence of h is better noticed when, instead of deriving the Cea estimate through the Galerkin projector G_h, it is obtained by analyzing individually each of the errors $\|\sigma - \sigma_h\|_H$ and $\|u - u_h\|_Q$. More precisely, we have the following theorem.

Theorem 2.6. *Under the same assumptions and notations of Theorems* 2.1 *and* 2.4, *there hold*

$$\|\sigma - \sigma_h\|_H \le \left(1 + \frac{\|\mathbf{A}\|}{\alpha_h}\right)\left(1 + \frac{\|\mathbf{B}\|}{\beta_h}\right) \inf_{\zeta_h \in H_h} \|\sigma - \zeta_h\|_H + \frac{\|\mathbf{B}\|}{\alpha_h} \inf_{w_h \in Q_h} \|u - w_h\|_Q$$

and

$$\|u - u_h\|_Q \le \frac{\|\mathbf{A}\|}{\beta_h}\left(1 + \frac{\|\mathbf{A}\|}{\alpha_h}\right)\left(1 + \frac{\|\mathbf{B}\|}{\beta_h}\right) \inf_{\zeta_h \in H_h} \|\sigma - \zeta_h\|_H$$

$$+ \quad \left(1 + \frac{\|\mathbf{B}\|}{\beta_h} + \frac{\|\mathbf{A}\|\,\|\mathbf{B}\|}{\alpha_h\,\beta_h}\right) \inf_{w_h \in Q_h} \|u - w_h\|_Q\,.$$

Proof. We begin by defining the set

$$V_h^g := \{\tau_h \in H_h : \quad b(\tau_h, v_h) = G(v_h) \quad \forall v_h \in Q_h\}$$

and observing, according to the second equation of (2.15), that $\sigma_h \in V_h^g$ and that $(\sigma_h - \tau_h^g) \in V_h \quad \forall \tau_h^g \in V_h^g$. Then, we aim to bound $\|\sigma - \sigma_h\|_H$ in terms of $\mathrm{dist}(u, Q_h)$ and $\mathrm{dist}(\sigma, V_h^g)$. To this end, we first apply the triangle inequality and obtain

$$\|\sigma - \sigma_h\|_H \le \|\sigma - \tau_h^g\|_H + \|\sigma_h - \tau_h^g\|_H \qquad \forall \tau_h^g \in V_h^g. \tag{2.80}$$

Then, using the discrete inf-sup condition (2.74), which is equivalent to hypothesis (i) of Theorem 2.4, we have

$$\alpha_h\,\|\sigma_h - \tau_h^g\|_H \le \sup_{\substack{\tau_h \in V_h \\ \tau_h \neq 0}} \frac{a(\sigma_h - \tau_h^g, \tau_h)}{\|\tau_h\|_H}. \tag{2.81}$$

Now, adding and subtracting σ, and using from the first equations of (2.1) and (2.71) that

$$a(\sigma - \sigma_h, \tau_h) + b(\tau_h, u - u_h) = 0 \qquad \forall \tau_h \in H_h, \tag{2.82}$$

and then adding and subtracting $w_h \in Q_h$, we deduce that for each $\tau_h \in V_h$ there holds

$$\begin{aligned} a(\sigma_h - \tau_h^g, \tau_h) &= a(\sigma_h - \sigma, \tau_h) + a(\sigma - \tau_h^g, \tau_h) \\ &= b(\tau_h, u - u_h) + a(\sigma - \tau_h^g, \tau_h) \\ &= b(\tau_h, u - w_h) + b(\tau_h, w_h - u_h) + a(\sigma - \tau_h^g, \tau_h) \\ &= b(\tau_h, u - w_h) + a(\sigma - \tau_h^g, \tau_h) \qquad \forall w_h \in Q_h. \end{aligned} \tag{2.83}$$

Note here that the term $b(\tau_h, w_h - u_h)$ vanishes since $\tau_h \in V_h$. Then, substituting the preceding expression into (2.81) and recalling from (2.2) and (2.3) that $\|\mathbf{A}\|$ and $\|\mathbf{B}\|$ are the boundedness constants for a and b, respectively, we obtain

$$\alpha_h \|\sigma_h - \tau_h^g\|_H \leq \sup_{\substack{\tau_h \in V_h \\ \tau_h \neq 0}} \frac{b(\tau_h, u - w_h) + a(\sigma - \tau_h^g, \tau_h)}{\|\tau_h\|_H}$$

$$\leq \|\mathbf{B}\| \, \|u - w_h\|_Q + \|\mathbf{A}\| \, \|\sigma - \tau_h^g\|_H \qquad \forall w_h \in Q_h,$$

which yields

$$\|\sigma_h - \tau_h^g\|_H \leq \frac{\|\mathbf{B}\|}{\alpha_h} \mathrm{dist}(u, Q_h) + \frac{\|\mathbf{A}\|}{\alpha_h} \|\sigma - \tau_h^g\|_H \qquad \forall \tau_h^g \in V_h^g.$$

Thus, (2.80) and the preceding estimate directly imply that

$$\|\sigma - \sigma_h\|_H \leq \left(1 + \frac{\|\mathbf{A}\|}{\alpha_h}\right) \mathrm{dist}(\sigma, V_h^g) + \frac{\|\mathbf{B}\|}{\alpha_h} \mathrm{dist}(\mathrm{u}, \mathrm{Q_h}). \tag{2.84}$$

Having established (2.84), it remains now to bound $\mathrm{dist}(\sigma, V_h^g)$ to complete the a priori estimate for $\|\sigma - \sigma_h\|_H$.

To this end, we note that the application of Lemma 2.1 to the present finite-dimensional setting implies that the discrete inf-sup condition for b given by hypothesis (ii) of Theorem 2.4 [cf. (2.72)] is equivalent, in particular, to the fact that $\mathbf{B}_h$ is a bijection from $V_h^\perp$ into Q_h, and there holds

$$\|\mathbf{B}_h(\tau_h)\|_Q \geq \beta_h \|\tau_h\|_H \qquad \forall \tau_h \in V_h^\perp. \tag{2.85}$$

Notice here that the orthogonality of V_h is with respect to the space H_h, that is, $H_h = V_h \bigoplus V_h^\perp$. Hence, given $\zeta_h \in H_h$, there exists a unique $\overline{\zeta}_h \in V_h^\perp$ such that $\mathbf{B}_h(\overline{\zeta}_h) = \mathbf{B}_h(\sigma_h - \zeta_h)$, that is,

$$b(\overline{\zeta}_h, w_h) = b(\sigma_h - \zeta_h, w_h) \qquad \forall w_h \in Q_h, \tag{2.86}$$

and, thanks to (2.85),

$$\|\overline{\zeta}_h\|_H \leq \frac{1}{\beta_h} \|\mathbf{B}_h(\sigma_h - \zeta_h)\|_Q. \tag{2.87}$$

Next, since $b(\sigma_h - \zeta_h, w_h) = b(\sigma - \zeta_h, w_h) \ \ \forall w_h \in Q_h$, which follows from the second equations of (2.1) and (2.71), we obtain

$$\|\mathbf{B}_h(\sigma_h - \zeta_h)\|_Q = \sup_{\substack{w_h \in Q_h \\ w_h \neq 0}} \frac{b(\sigma_h - \zeta_h, w_h)}{\|w_h\|_Q}$$

$$= \sup_{\substack{w_h \in Q_h \\ w_h \neq 0}} \frac{b(\sigma - \zeta_h, w_h)}{\|w_h\|_Q} \leq \|\mathbf{B}\| \, \|\sigma - \zeta_h\|_H,$$

which, inserted back into (2.87), gives

$$\|\overline{\zeta}_h\|_H \le \frac{\|\mathbf{B}\|}{\beta_h}\,\|\sigma - \zeta_h\|_H\,. \tag{2.88}$$

Thus, it is clear from (2.86) and the second equation of the Galerkin scheme (2.71) that $\zeta_h + \overline{\zeta}_h \in V_h^g$, and therefore, using also (2.88), we find that

$$\begin{aligned}\operatorname{dist}(\sigma, V_h^g) &\le \|\sigma - (\zeta_h + \overline{\zeta}_h)\|_H \\ &\le \|\sigma - \zeta_h\|_H + \|\overline{\zeta}_h\|_H \\ &\le \left(1 + \frac{\|\mathbf{B}\|}{\beta_h}\right)\|\sigma - \zeta_h\|_H \qquad \forall\, \zeta_h \in H_h,\end{aligned}$$

that is,

$$\operatorname{dist}(\sigma, V_h^g) \le \left(1 + \frac{\|\mathbf{B}\|}{\beta_h}\right)\operatorname{dist}(\sigma, H_h). \tag{2.89}$$

In this way, (2.84) and (2.89) complete the Cea estimate for $\|\sigma - \sigma_h\|_H$.

Furthermore, applying again the triangle inequality we have

$$\|u - u_h\|_Q \le \|u - w_h\|_Q + \|u_h - w_h\|_Q \quad \forall w_h \in Q_h. \tag{2.90}$$

Hence, employing the discrete inf-sup condition for b [cf. (2.72)], subtracting and adding u, and utilizing the identity (2.82), we obtain

$$\begin{aligned}\beta_h\|u_h - w_h\|_Q &\le \sup_{\substack{\tau_h \in H_h\\ \tau_h \neq 0}} \frac{b(\tau_h, u_h - w_h)}{\|\tau_h\|_H} \\ &= \sup_{\substack{\tau_h \in H_h\\ \tau_h \neq 0}} \frac{b(\tau_h, u_h - u) + b(\tau_h, u - w_h)}{\|\tau_h\|_H} \\ &= \sup_{\substack{\tau_h \in H_h\\ \tau_h \neq 0}} \frac{a(\sigma - \sigma_h, \tau_h) + b(\tau_h, u - w_h)}{\|\tau_h\|_H} \\ &\le \|\mathbf{A}\|\,\|\sigma - \sigma_h\|_H + \|\mathbf{B}\|\,\|u - w_h\|_Q \quad \forall w_h \in Q_h,\end{aligned}$$

and inserting the preceding estimate in (2.90), we conclude that

$$\|u - u_h\|_Q \le \left(1 + \frac{\|\mathbf{B}\|}{\beta_h}\right)\operatorname{dist}(u, Q_h) + \frac{\|\mathbf{A}\|}{\beta_h}\|\sigma - \sigma_h\|_H. \tag{2.91}$$

Finally, (2.91) and the a priori bound for the error $\|\sigma - \sigma_h\|_H$ imply the Cea estimate for $\|u - u_h\|_Q$, thus completing the proof.

□

It is interesting to observe at this point that if $V_h \subseteq V$, then the expression $b(\tau_h, u - u_h)$ vanishes $\forall\, \tau_h \in V_h$, and hence it is not necessary to add and subtract $w_h \in Q_h$ in (2.83). Indeed, in that case (2.84) simply reduces to

$$\|\sigma - \sigma_h\|_H \leq \left(1 + \frac{\|\mathbf{A}\|}{\alpha_h}\right) \operatorname{dist}(\sigma, V_h^g),$$

which, together with (2.89), gives

$$\|\sigma - \sigma_h\|_H \leq \left(1 + \frac{\|\mathbf{A}\|}{\alpha_h}\right) \left(1 + \frac{\|\mathbf{B}\|}{\beta_h}\right) \operatorname{dist}(\sigma, H_h).$$

In other words, when the discrete kernel V_h of b is contained in the continuous kernel V, the Cea estimate for $\|\sigma - \sigma_h\|_H$ depends only on $\operatorname{dist}(\sigma, H_h)$ but not on $\operatorname{dist}(u, Q_h)$.

On the other hand, it is also important to remark that the discrete inf-sup condition for b is fundamental for the uniqueness of u_h. In fact, analogously to the continuous case, we know that this hypothesis is rewritten as

$$\|\mathbf{B}_h^*(v_h)\|_H \geq \beta_h \|v_h\|_Q \quad \forall v_h \in Q_h,$$

which, in the present discrete case, is equivalent to the injectivity of $\mathbf{B}_h^*$. Therefore, if it is not satisfied, then there must exist $w \in Q_h$, $w \neq 0$, such that $\mathbf{B}_h^*(w) = 0$, and hence, given any solution $(\sigma_h, u_h) \in H_h \times Q_h$ of the Galerkin scheme (2.71), we have that for each $c \in \mathbb{R}$, $(\sigma_h, u_h + cw)$ is solution of (2.71) as well.

We conclude this section by emphasizing that the subspaces H_h and Q_h defining the Galerkin scheme (2.71) cannot be chosen arbitrarily since, obviously, they need to satisfy the hypotheses from Theorem 2.4. In fact, the most demanding of all is the discrete inf-sup condition for b [cf. (2.72)]. In particular, since it is equivalent to the surjectivity of $\mathbf{B}_h : H_h \to Q_h$, we deduce that a necessary condition for its occurrence is that $\dim H_h \geq \dim Q_h$. Then, the following result, which is known as Fortin's trick and has been employed in diverse applications (see, e.g., Sect. 4.2), provides a sufficient condition for it. More precisely, the following lemma establishes that the continuous inf-sup condition for b, together with the existence of a suitable sequence of uniformly bounded operators, called FORTIN'S OPERATORS, implies the corresponding discrete inf-sup condition for b.

Lemma 2.6 (Fortin's Lemma). *Let H and Q be Hilbert spaces, let $b : H \times Q \to \mathbb{R}$ be a bounded bilinear form, and assume that there exists $\beta > 0$ such that*

$$\sup_{\substack{\tau \in H \\ \tau \neq 0}} \frac{b(\tau, v)}{\|\tau\|_H} \geq \beta \|v\|_Q \quad \forall v \in Q.$$

In addition, let $\{H_h\}_{h \in I}$ and $\{Q_h\}_{h \in I}$ be sequences of subspaces of H and Q, respectively, and assume that there exist $\{\Pi_h\}_{h \in I} \subseteq \mathscr{L}(H, H_h)$ and $\tilde{C} > 0$ such that

$$\|\Pi_h\| \leq \tilde{C} \qquad \forall h \in I$$

and

$$b(\tau - \Pi_h(\tau), v_h) = 0 \qquad \forall \tau \in H, \qquad \forall v_h \in Q_h, \quad \forall h \in I.$$

Then there exists $\tilde{\beta} > 0$, independently of h, such that

$$\sup_{\substack{\tau_h \in H_h \\ \tau_h \neq 0}} \frac{b(\tau_h, v_h)}{\|\tau_h\|_H} \geq \tilde{\beta}\, \|v_h\|_Q \qquad \forall v_h \in Q_h, \quad \forall h \in I.$$

Proof. Given $v_h \in Q_h$, the continuous inf-sup condition for b and the remaining hypotheses yield

$$\begin{aligned} \beta \|v_h\|_Q &\leq \sup_{\substack{\tau \in H \\ \tau \neq 0}} \frac{|b(\tau, v_h)|}{\|\tau\|_H} = \sup_{\substack{\tau \in H \\ \tau \neq 0}} \frac{|b(\Pi_h(\tau), v_h)|}{\|\tau\|_H} \\ &\leq \tilde{C} \sup_{\substack{\tau \in H \\ \Pi_h(\tau) \neq 0}} \frac{|b(\Pi_h(\tau), v_h)|}{\|\Pi_h(\tau)\|_H} \\ &\leq \tilde{C} \sup_{\substack{\tau_h \in H_h \\ \tau_h \neq 0}} \frac{b(\tau_h, v_h)}{\|\tau_h\|_H}, \end{aligned}$$

which shows the required inequality with $\tilde{\beta} := \beta / \tilde{C}$.

□

Chapter 3
Raviart-Thomas Spaces

In this chapter we introduce Raviart–Thomas spaces, which constitute the most classical finite element subspaces of $H(\text{div};\Omega)$, and prove their main interpolation and approximation properties. Several aspects of our analysis follow the approaches from [16, 50, 52].

3.1 Preliminary Results

In what follows, Ω is a bounded and connected domain of $\mathbb{R}^n$, $n \in \{2,3\}$, with polyhedral boundary Γ, and $\mathscr{T}_h$ is a triangularization of $\overline{\Omega}$. More precisely, $\mathscr{T}_h$ is a finite family of triangles (in $\mathbb{R}^2$) or tetrahedra (in $\mathbb{R}^3$), such that

(i) $\overline{\Omega} = \bigcup_{K \in \mathscr{T}_h} K$;

(ii) $\overset{\circ}{K} \neq 0 \quad \forall K \in \mathscr{T}_h$;

(iii) $\overset{\circ}{K_i} \cap \overset{\circ}{K_j} = \phi \quad \forall K_i, K_j \in \mathscr{T}_h,\ K_i \neq K_j$;

(iv) If $F = K_i \cap K_j,\ K_i, K_j \in \mathscr{T}_h,\ K_i \neq K_j$, then F is a common face, a common side, or a common vertex of K_i and K_j;

(v) $\text{diam}(K) =: h_K \leq h \qquad \forall K \in \mathscr{T}_h$.

In addition, to each $\mathscr{T}_h$ we associate a fixed reference polyhedron $\hat{K}$, which can or cannot belong to $\mathscr{T}_h$, and a family of affine mappings $\{T_K\}_{K \in \mathscr{T}_h}$ such that

(a) $T_K : \mathbb{R}^n \to \mathbb{R}^n, T_K(\hat{x}) = B_K \hat{x} + b_K \ \ \forall \hat{x} \in \mathbb{R}^n$, with $B_K \in \mathbb{R}^{n \times n}$ invertible, and $b_K \in \mathbb{R}^n$;

(b) $K = T_K(\hat{K}) \quad \forall K \in \mathscr{T}_h$.

One usually considers $\hat{K}$ as the unit simplex, that is, the triangle with vertices (1,0), (0,1), and (0,0) in $\mathbb{R}^2$, or the tetrahedron with vertices (1,0,0), (0,1,0), (0,0,1), and (0,0,0) in $\mathbb{R}^3$.

Throughout the rest of this section we demonstrate a sequence of results characterizing the spaces $H^1(\Omega)$ and $H(\text{div};\Omega)$ in terms of their local behaviors on the

G.N. Gatica, *A Simple Introduction to the Mixed Finite Element Method: Theory and Applications*, SpringerBriefs in Mathematics, DOI 10.1007/978-3-319-03695-3_3,

elements of the triangularization $\mathcal{T}_h$. In what follows, $\langle\cdot,\cdot\rangle_{\partial K}$ denotes the duality between $H^{-1/2}(\partial K)$ and $H^{1/2}(\partial K)$ for each $K \in \mathcal{T}_h$. Next, we omit the symbol $\gamma_{\mathbf{n}}$ to denote the respective normal traces and simply write, when no confusion arises, $\tau\cdot\mathbf{n} \quad \forall\tau \in H(\mathrm{div};\Omega)$ and $\tau\cdot\mathbf{n}_K \quad \forall\tau\in H(\mathrm{div};K)$, where $\mathbf{n}_K$ is the normal vector to ∂K. Similarly, we omit the symbol γ_0 and just write $v|_\Gamma$ (or only v) for $v\in H^1(\Omega)$ and $v|_{\partial K}$ (or only v) for $v \in H^1(K)$.

Lemma 3.1. *Define the spaces* $X := \left\{ v\in L^2(\Omega): \quad v|_K \in H^1(K) \quad \forall K\in\mathcal{T}_h \right\}$ *and* $H_0(\mathrm{div};\Omega) := \left\{\tau\in H(\mathrm{div};\Omega): \quad \tau\cdot\mathbf{n} = 0 \text{ on } \Gamma\right\}$. *Then*

$$H^1(\Omega) = \left\{ v\in X: \quad \sum_{K\in\mathcal{T}_h} \langle\tau\cdot\mathbf{n}_K, v\rangle_{\partial K} = 0 \quad \forall\tau\in H_0(\mathrm{div};\Omega)\right\}.$$

Proof. We proceed by double inclusion. Let $v\in X$ such that

$$\sum_{K\in\mathcal{T}_h} \langle\tau\cdot\mathbf{n}_K, v\rangle_{\partial K} = 0 \quad \forall\tau\in H_0(\mathrm{div};\Omega).$$

Since $v|_K\in H^1(K) \quad \forall K\in\mathcal{T}_h$, we have for each $\tau\in H_0(\mathrm{div};\Omega)$ that

$$\int_K \tau\cdot\nabla v = -\int_K v\,\mathrm{div}\,\tau + \langle\tau\cdot\mathbf{n}_K, v\rangle_{\partial K},$$

which yields

$$\sum_{K\in\mathcal{T}_h}\int_K \tau\cdot\nabla v = -\int_\Omega v\,\mathrm{div}\,\tau.$$

In particular, for $\tau\in[C_0^\infty(\Omega)]^n \subseteq H_0(\mathrm{div};\Omega)$ the preceding identity becomes

$$\langle\nabla v,\tau\rangle_{[\mathcal{D}'(\Omega)]^n\times[\mathcal{D}(\Omega)]^n} = \sum_{K\in\mathcal{T}_h}\int_K\tau\cdot\nabla v = \int_\Omega \tau\cdot w,$$

where $\langle\cdot,\cdot\rangle_{[\mathcal{D}'(\Omega)]^n\times[\mathcal{D}(\Omega)]^n}$ stands for the distributional pairing of $[\mathcal{D}'(\Omega)]^n$ and $[\mathcal{D}(\Omega)]^n$, and $w\in[L^2(\Omega)]^n$ is given by $w|_K = \nabla(v|_K) \quad \forall K\in\mathcal{T}_h$. This proves that $\nabla v = w$ in $[\mathcal{D}'(\Omega)]^n$, and hence $v\in H^1(\Omega)$.

Conversely, let $v\in H^1(\Omega)$. It is clear that $v\in X$ since obviously $v\in L^2(\Omega)$ and $v|_K\in H^1(K) \quad \forall K\in\mathcal{T}_h$. Now, given $\tau\in H_0(\mathrm{div};\Omega)$, we utilize the Green identity (1.50) (cf. Lemma 1.4) in $H(\mathrm{div};\Omega)$ and $H(\mathrm{div};K) \quad \forall K\in\mathcal{T}_h$ to deduce that

$$\begin{aligned} 0 = \langle\tau\cdot\mathbf{n}, v\rangle_\Gamma &= \int_\Omega\tau\cdot\nabla v + \int_\Omega v\,\mathrm{div}\,\tau \\ &= \sum_{K\in\mathcal{T}_h}\int_K\tau\cdot\nabla v + \int_\Omega v\,\mathrm{div}\,\tau \\ &= \sum_{K\in\mathcal{T}_h}\left\{-\int_K v\,\mathrm{div}\,\tau + \langle\tau\cdot\mathbf{n}_K, v\rangle_{\partial K}\right\} + \int_\Omega v\,\mathrm{div}\,\tau \end{aligned}$$

$$= \sum_{K\in\mathscr{T}_h} \langle \tau\cdot\mathbf{n}_K, v\rangle_{\partial K},$$

which completes the proof.

□

An immediate consequence of the preceding theorem is given by the following result.

Lemma 3.2. *Let* $X := \left\{ v \in L^2(\Omega) : \quad v|_K \in H^1(K) \quad \forall K \in \mathscr{T}_h \right\}$. *Then*

$$H^1(\Omega) = \left\{ v \in X : \quad \sum_{K\in\mathscr{T}_h} \int_{\partial K} \tau\cdot\mathbf{n}_K v = 0 \quad \forall \tau \in [C_0^\infty(\Omega)]^n \right\}.$$

Proof. The proof follows by employing Lemma 3.1, the inclusion $[C_0^\infty(\Omega)]^n \subseteq H_0(\mathrm{div};\Omega)$, the fact that $\tau|_K \in [H^1(K)]^n \quad \forall \tau \in [C_0^\infty(\Omega)]^n, \quad \forall K \in \mathscr{T}_h$, and the identity [cf. (1.45)]

$$\langle \tau\cdot\mathbf{n}_K, v\rangle_{\partial K} = \int_{\partial K} \tau\cdot\mathbf{n}_K v \quad \forall v \in H^1(K), \quad \forall \tau \in [H^1(K)]^n.$$

We omit further details.

□

To further simplify the characterization of $H^1(\Omega)$ given by the previous lemmas, we need the following technical result.

Lemma 3.3. *Let* $K_i, K_j \in \mathscr{T}_h$ *be adjacent polyhedra with common face/side* F, *and let* $z \in L^2(F)$ *such that* $\int_F z\rho = 0 \quad \forall \rho \in C_0^\infty(K_i \cup K_j)$. *Then* $z = 0$ *on* F.

Proof. Using that $C_0^\infty(F)$ is dense in $L^2(F)$, it suffices to show that $\int_F z\varphi = 0 \quad \forall \varphi \in C_0^\infty(F)$. To this end, let G be a line perpendicular to F, and let $x = (x_1, x_2, \cdots, x_n)$ be the representation of a coordinate system with $(x_1, x_2, \cdots, x_{n-1}) \in F$, $x_n \in G$, and the origin given by the intersection point of F and G (which can be assumed to be the barycenter of F). Then, given $\varphi \in C_0^\infty(F)$, we can construct, via regularization techniques, a function $\psi \in C_0^\infty(G)$ such that $\psi(0) = 1$, and so that $\operatorname{sop}\varphi \times \operatorname{sop}\psi$ is contained in the interior of $K_i \cup K_j$. Hence, defining the function $\rho(x) := \varphi(x_1, x_2, \cdots, x_{n-1})\,\psi(x_n)$, we have that $\rho \in C_0^\infty(K_i \cup K_j)$ and $\rho|_F = \varphi$, which implies that $0 = \int_F z\rho = \int_F z\varphi$, thus completing the proof.

□

We are now able to prove the following theorem.

Theorem 3.1. *Let* $X := \left\{ v \in L^2(\Omega) : \quad v|_K \in H^1(K) \quad \forall K \in \mathscr{T}_h \right\}$. *Then*

$$H^1(\Omega) = \Big\{ v \in X : \quad v|_{K_i} - v|_{K_j} = 0 \quad \textit{in} \quad L^2(F)$$
$$\forall K_i, K_j \in \mathscr{T}_h \quad \textit{that are adjacent with common face/side } F \Big\}.$$

Proof. Let $v \in X$ such that $v|_{K_i} - v|_{K_j} = 0$ in $L^2(F)$ $\forall K_i, K_j \in \mathscr{T}_h$ that are adjacent with common face/side F. Then, given $\tau \in [C_0^\infty(\Omega)]^n$, we have $\tau \cdot \mathbf{n} = 0$ in Γ, and hence

$$\sum_{K \in \mathscr{T}_h} \int_{\partial K} \tau \cdot \mathbf{n}_K v = \sum_{F \in I_h(\Omega)} \int_F \left(v|_{K_{i,F}} - v|_{K_{j,F}} \right) \tau \cdot \mathbf{n}_{K_{i,F}},$$

where $I_h(\Omega)$ is the set of interior faces/sides of $\mathscr{T}_h$, and $K_{i,F}$ and $K_{j,F}$ are the adjacent polyhedra with common face/side F. Note here that $\mathbf{n}_{K_{i,F}} = -\mathbf{n}_{K_{j,F}}$. It follows that

$$\sum_{K \in \mathscr{T}_h} \int_{\partial K} \tau \cdot \mathbf{n}_K v = 0 \qquad \forall \tau \in [C_0^\infty(\Omega)]^n,$$

which, thanks to Lemma 3.2, implies that $v \in H^1(\Omega)$.

Conversely, let $v \in H^1(\Omega)$. It is clear from Lemma 3.2 that

$$\sum_{K \in \mathscr{T}_h} \int_{\partial K} \tau \cdot \mathbf{n}_K v = 0 \qquad \forall \tau \in [C_0^\infty(\Omega)]^n.$$

In particular, given $\tau \in [C_0^\infty(K_i \cup K_j)]^n$, with $K_i, K_j \in \mathscr{T}_h$ adjacent with common face/side F, we obtain

$$\begin{aligned} 0 = \sum_{K \in \mathscr{T}_h} \int_{\partial K} \tau \cdot \mathbf{n}_K v &= \int_F \left(v|_{K_i} - v|_{K_j} \right) \tau \cdot \mathbf{n}_{K_i} \\ &= \int_F \left(v|_{K_i} - v|_{K_j} \right) \mathbf{n}_{K_i} \cdot \tau, \end{aligned}$$

whence, applying Lemma 3.3 to a nonnull component of $\mathbf{n}_{K_i}$, we deduce that $v|_{K_i} - v|_{K_j} = 0$ in $L^2(F)$.

□

Our next goal is to characterize the space $H(\text{div};\Omega)$ in terms of the local behaviors. We begin with the following lemma, which constitutes a kind of dual result to Lemma 3.1.

Lemma 3.4. *Let* $Y := \left\{ \tau \in [L^2(\Omega)]^n : \quad \tau|_K \in H(\text{div}; K) \quad \forall K \in \mathscr{T}_h \right\}$. *Then*

$$H(\text{div};\Omega) = \left\{ \tau \in Y : \quad \sum_{K \in \mathscr{T}_h} \langle \tau \cdot \mathbf{n}_K, v \rangle_{\partial K} = 0 \quad \forall v \in H_0^1(\Omega) \right\}.$$

Proof. We proceed by double inclusion. Let $\tau \in Y$ such that

$$\sum_{K \in \mathscr{T}_h} \langle \tau \cdot \mathbf{n}_K, v \rangle_{\partial K} = 0 \qquad \forall v \in H_0^1(\Omega).$$

Since $\tau|_K \in H(\text{div};K) \quad \forall K \in \mathscr{T}_h$, we have for each $v \in H_0^1(\Omega)$

$$\int_K v \operatorname{div} \tau = -\int_K \tau \cdot \nabla v + \langle \tau \cdot \mathbf{n}_K, v \rangle_{\partial K},$$

which gives

$$\sum_{K \in \mathscr{T}_h} \int_K v \operatorname{div} \tau = -\int_\Omega \tau \cdot \nabla v.$$

In particular, for $v \in C_0^\infty(\Omega) \subseteq H_0^1(\Omega)$ the preceding identity reduces to

$$\langle \operatorname{div} \tau, v \rangle_{\mathscr{D}'(\Omega) \times \mathscr{D}(\Omega)} = \sum_{K \in \mathscr{T}_h} \int_K v \operatorname{div} \tau = \int_\Omega v z,$$

where $\langle \cdot, \cdot \rangle_{\mathscr{D}'(\Omega) \times \mathscr{D}(\Omega)}$ is the distributional pairing of $\mathscr{D}'(\Omega)$ and $\mathscr{D}(\Omega)$, and $z \in L^2(\Omega)$ is given by $z|_K = \operatorname{div}(\tau|_K) \quad \forall K \in \mathscr{T}_h$. This shows that $\operatorname{div} \tau = z$ in $\mathscr{D}'(\Omega)$, and hence $\tau \in H(\text{div};\Omega)$.

Conversely, let $\tau \in H(\text{div};\Omega)$. It is clear that $\tau \in Y$ since obviously $\tau \in [L^2(\Omega)]^n$ and $\tau|_K \in H(\text{div};K) \quad \forall K \in \mathscr{T}_h$. Thus, given $v \in H_0^1(\Omega)$, we first utilize the Green identity (1.50) (cf. Lemma 1.4) in $H(\text{div};\Omega)$ and $H(\text{div};K) \quad \forall K \in \mathscr{T}_h$ and proceed as in the second part of the proof of Lemma 3.1 to conclude that

$$0 = \langle \tau \cdot \mathbf{n}, v \rangle_\Gamma = \sum_{K \in \mathscr{T}_h} \langle \tau \cdot \mathbf{n}_K, v \rangle_{\partial K},$$

which completes the proof.

□

The following theorem is a consequence of the preceding lemma and the technical result given by Lemma 3.3.

Theorem 3.2. *Let* $Z := \left\{ \tau \in [L^2(\Omega)]^n : \quad \tau|_K \in [H^1(K)]^n \quad \forall K \in \mathscr{T}_h \right\}$. *Then*

$$H(\text{div};\Omega) \cap Z = \Big\{ \tau \in Z: \quad \tau \cdot \mathbf{n}_{K_i} + \tau \cdot \mathbf{n}_{K_j} = 0 \quad in \quad L^2(F)$$
$$\forall K_i, K_j \in \mathscr{T}_h \quad \text{that are adjacent with common face/side } F \Big\}.$$

Proof. Let $\tau \in Z$ such that $\tau \cdot \mathbf{n}_{K_i} + \tau \cdot \mathbf{n}_{K_j} = 0$ in $L^2(F) \quad \forall K_i, K_j \in \mathscr{T}_h$ that are adjacent with common face/side F. Then, given $v \in H_0^1(\Omega)$, we use that $\tau \cdot \mathbf{n}_K \in L^2(\partial K)$, since $\tau|_K \in [H^1(K)]^n \quad \forall K \in \mathscr{T}_h$, and employ the same notation of Theorem 3.1 to deduce that

$$\sum_{K \in \mathscr{T}_h} \langle \tau \cdot \mathbf{n}_K, v \rangle_{\partial K} = \sum_{K \in \mathscr{T}_h} \int_{\partial K} \tau \cdot \mathbf{n}_K v = \sum_{F \in I_h(\Omega)} \int_F \left(\tau \cdot \mathbf{n}_{K_{i,F}} + \tau \cdot \mathbf{n}_{K_{j,F}} \right) v = 0,$$

which, thanks to Lemma 3.4, yields $\tau \in H(\text{div};\Omega)$.

Conversely, let $\tau \in H(\text{div};\Omega) \cap Z$. It follows again from Lemma 3.4 that

$$\sum_{K\in\mathcal{T}_h} \langle \tau \cdot \mathbf{n}_K, v \rangle_{\partial K} = 0 \qquad \forall v \in H_0^1(\Omega).$$

In particular, for $v \in C_0^\infty(K_i \cup K_j)$, where $K_i, K_j \in \mathcal{T}_h$ are adjacent polyhedra with common face/side F, we find that

$$0 = \sum_{K\in\mathcal{T}_h} \langle \tau \cdot \mathbf{n}_K, v \rangle_{\partial K} = \sum_{K\in\mathcal{T}_h} \int_{\partial K} \tau \cdot \mathbf{n}_K \, v = \int_F \left(\tau \cdot \mathbf{n}_{K_i} + \tau \cdot \mathbf{n}_{K_j} \right) v,$$

and hence, in virtue of Lemma 3.3, we conclude that $\tau \cdot \mathbf{n}_{K_i} + \tau \cdot \mathbf{n}_{K_j} = 0$ in $L^2(F)$. □

3.2 Spaces of Polynomials

Given a bounded and convex domain S of $\mathbb{R}^n, n \in \{2,3\}$ and a nonnegative integer k, we define the spaces

$$\tilde{\mathbb{P}}_k(S) := \{p : S \to \mathbb{R} : \quad p \text{ is a polynomial of degree} = k\}$$

and

$$\mathbb{P}_k(S) := \{p : S \to \mathbb{R} : \quad p \text{ is a polynomial of degree} \leq k\}.$$

Equivalently, denoting $\mathbb{N}_0 := \mathbb{N} \cup \{0\}$ and using a multi-index notation, we have that $p \in \tilde{\mathbb{P}}_k(S)$ if and only if there exist scalars $a_\alpha \in \mathbb{R}$ for all $\alpha := (\alpha_1, \alpha_2, \cdots, \alpha_n)$ $\in \mathbb{N}_0^n$ with $|\alpha| := \sum_{j=1}^{n} \alpha_j = k$ such that

$$p(x) = \sum_{|\alpha|=k} a_\alpha x^\alpha \qquad \forall x \in S,$$

where $x^\alpha := x_1^{\alpha_1} x_2^{\alpha_2}, \cdots x_n^{\alpha_n}$. Analogously, $p \in \mathbb{P}_k(S)$ if and only if there exist scalars $a_\alpha \in \mathbb{R}$ for all $\alpha := (\alpha_1, \alpha_2, \cdots, \alpha_n) \in \mathbb{N}_0^n$ with $|\alpha| \leq k$ such that

$$p(x) = \sum_{|\alpha|\leq k} a_\alpha x^\alpha \qquad \forall x \in S.$$

It can be proved that $\dim \tilde{\mathbb{P}}_k(S) = \binom{n+k-1}{k}$, and hence, using that

$$\binom{n}{j} + \binom{n}{j+1} = \binom{n+1}{j+1} \qquad \forall j \in \mathbb{N}_0,$$

we deduce that

$$\dim \mathbb{P}_k(S) = \sum_{j=0}^{k} \dim \tilde{\mathbb{P}}_j(S) = \sum_{j=0}^{k} \binom{n+j-1}{j} = \binom{n+k}{k}. \tag{3.1}$$

Furthermore, we define the Raviart–Thomas space of order $k \geq 0$ on S as

$$RT_k(S) := [\mathbb{P}_k(S)]^n + \mathbb{P}_k(S)x, \tag{3.2}$$

that is, $\mathbf{p} \in RT_k(S)$ if and only if there exist $p_0, p_1, \cdots, p_n \in \mathbb{P}_k(S)$ such that

$$\mathbf{p}(x) = \begin{pmatrix} p_1(x) \\ p_2(x) \\ \vdots \\ p_n(x) \end{pmatrix} + p_0(x) \begin{pmatrix} x_1 \\ x_2 \\ \vdots \\ x_n \end{pmatrix} \quad \forall x := (x_1, x_2, \cdots, x_n)^{\mathrm{t}} \in S. \tag{3.3}$$

Lemma 3.5. *There holds* $RT_k(S) = [\mathbb{P}_k(S)]^n \oplus \tilde{\mathbb{P}}_k(S)\, x$ *and*

$$\dim RT_k(S) = \frac{(n+k+1)\,(n+k-1)!}{(n-1)!\; k!}.$$

Proof. Since $\tilde{\mathbb{P}}_k(S) \subseteq \mathbb{P}_k(S)$, it is clear that $[\mathbb{P}_k(S)]^n \oplus \tilde{\mathbb{P}}_k(S)x \subseteq RT_k(S)$. For the converse inclusion we let $\mathbf{p} \in RT_k(S)$, as indicated in (3.3), and for each $i \in \{0,1,2,\cdots,n\}$ we let $a_\alpha^i \in \mathbb{R}$, $\forall \alpha \in \mathbb{N}_0^n$, with $|\alpha| \leq k$, such that

$$p_i(x) = \sum_{|\alpha| \leq k} a_\alpha^i x^\alpha \qquad \forall x \in S.$$

Hence, given $i \in \{1,2,\cdots,n\}$, it follows that the ith component of $\mathbf{p}$ is given by

$$\begin{aligned} p_i(x) + p_0(x)\, x_i &= \sum_{|\alpha| \leq k} a_\alpha^i\, x^\alpha + x_i \sum_{|\alpha| \leq k-1} a_\alpha^0\, x^\alpha + x_i \sum_{|\alpha| = k} a_\alpha^0\, x^\alpha \\ &= q_i(x) + x_i\, q_0(x) \qquad \forall x \in S, \end{aligned}$$

where

$$q_i(x) := \sum_{|\alpha| \leq k} a_\alpha^i\, x^\alpha + x_i \sum_{|\alpha| \leq k-1} a_\alpha^0\, x^\alpha \qquad \forall x \in S$$

and

$$q_0(x) := \sum_{|\alpha| = k} a_\alpha^0\, x^\alpha \qquad \forall x \in S.$$

Since $q_i \in \mathbb{P}_k(S) \quad \forall i \in \{1,2,\cdots,n\}$ and $q_0 \in \tilde{\mathbb{P}}_k(S)$, we deduce that $\mathbf{p}$ belongs to $[\mathbb{P}_k(S)]^n \oplus \tilde{\mathbb{P}}_k(S)x$, which completes the first identity of the lemma.

Finally, according to the preceding discussion, we have $\dim RT_k(S) = n\dim \mathbb{P}_k(S) + \dim \tilde{\mathbb{P}}_k(S)$, that is,

$$\dim RT_k(S) = n\binom{n+k}{k} + \binom{n+k-1}{k},$$

which, after simple algebraic computations, gives the stated formula. □

3.3 Local Raviart–Thomas Spaces

Let us consider again the triangularization $\mathscr{T}_h$ of $\overline{\Omega}$ introduced in Sect. 3.1. In what follows, we focus on the local Raviart–Thomas spaces $RT_k(K)\ \forall K \in \mathscr{T}_h,\ \forall k \geq 0$, whose elements are denoted from now on by $\boldsymbol{\tau}$ instead of $\mathbf{p}$.

Lemma 3.6. *For each $K \in \mathscr{T}_h$ there holds:*

(i) $\operatorname{div}\boldsymbol{\tau} \in \mathbb{P}_k(K) \quad \forall \boldsymbol{\tau} \in RT_k(K)$;
(ii) $\boldsymbol{\tau}\cdot\mathbf{n}_K|_F \in \mathbb{P}_k(F) \quad \forall$ *face/side F of K*, $\forall \boldsymbol{\tau} \in RT_k(K)$.

Proof. Given $K \in \mathscr{T}_h$ and $\boldsymbol{\tau} \in RT_k(K)$, we know from (3.3) that there exist polynomials $p_0, p_1, \cdots, p_n \in \mathbb{P}_k(K)$ such that τ_i, the ith component of $\boldsymbol{\tau}$, $i \in \{1,2,\cdots,n\}$, is given by

$$\tau_i(x) := p_i(x) + x_i\,p_0(x) \qquad \forall x \in K.$$

It follows easily that

$$(\operatorname{div}\boldsymbol{\tau})(x) = \sum_{i=1}^{n}\left\{\frac{\partial p_i(x)}{\partial x_i} + p_0(x) + x_i\frac{\partial p_0(x)}{\partial x_i}\right\},$$

which confirms the assertion (i).

On the other hand, let F be a side of the triangle K (in $\mathbb{R}^2$) or a face of the tetrahedron K (in $\mathbb{R}^3$). Then there exist scalars $a_1, a_2, \cdots, a_n, b \in \mathbb{R}$ such that F is contained in the line/plane of equation

$$a_1x_1 + a_2x_2 + \cdots + a_nx_n = b,$$

and hence the normal vector to F is given by $\mathbf{n}_K = \dfrac{\mathbf{a}}{\|\mathbf{a}\|}$, with $\mathbf{a} := (a_1, a_2, \cdots, a_n)^{\mathtt{t}}$. It follows that for each $x \in F$ there holds

$$\begin{aligned}
\left(\boldsymbol{\tau}\cdot\mathbf{n}_K\right)(\mathbf{x}) &= \sum_{i=1}^{n}\left\{p_i(x) + x_i\,p_0(x)\right\}\frac{a_i}{\|\mathbf{a}\|}\\
&= \frac{1}{\|\mathbf{a}\|}\left\{\sum_{i=1}^{n} a_i\,p_i(x) + p_0(x)\sum_{i=1}^{n} a_i x_i\right\}\\
&= \frac{1}{\|\mathbf{a}\|}\left\{\sum_{i=1}^{n} a_i\,p_i(x) + b\,p_0(x)\right\},
\end{aligned}$$

which proves (ii). □

Now, since F is contained in a hyperplane of dimension $n-1$, we obtain from formula (3.1) that

$$\dim \mathbb{P}_k(F) = \binom{n-1+k}{k} = \frac{(n+k-1)!}{(n-1)!\,k!} =: d_k\,.$$

Therefore, part (ii) of Lemma 3.6 guarantees that, given $K \in \mathscr{T}_h$, $\tau \in RT_k(K)$, and a face/side F of K, the normal component $\tau \cdot \mathbf{n}_K|_F$ is uniquely determined by

(a) The values of $\int_F \tau \cdot \mathbf{n}_K\, \psi \quad \forall\, \psi \in \mathbb{P}_k(F)$, or, equivalently,

(b) The values of $\int_F \tau \cdot \mathbf{n}_K\, \psi_j \quad \forall\, j \in \{1,2,\cdots,d_k\}$, where $\{\psi_1,\psi_2,\cdots,\psi_{d_k}\}$ is a basis of $\mathbb{P}_k(F)$.

In particular, for $k=0$ we have $d_0 = 1$, whereas

$$\dim RT_0(K) = \frac{(n+0+1)(n+0-1)!}{(n-1)!\,0!} = n+1,$$

which says that the dimension of $RT_0(K)$ coincides with the number of degrees of freedom generated by the normal components on the faces/sides F of K (three when K is a triangle of $\mathbb{R}^2$ and four when K is a tetrahedron of $\mathbb{R}^3$). However, we observe next that this coincidence does not hold for $k \geq 1$. In fact, the number of degrees of freedom on the $n+1$ faces/sides of K is given by $(n+1)d_k = \dfrac{(n+1)(n+k-1)!}{(n-1)!\,k!}$, whereas $\dim RT_k(K) = \dfrac{(n+k+1)(n+k-1)!}{(n-1)!\,k!}$, which indicates that additional degrees of freedom (most likely on K) need to be defined for an eventual unisolvency of the polynomials in $RT_k(K)$. The explicit result in this direction is provided by the following theorem.

Theorem 3.3. *Let $K \in \mathscr{T}_h$ and $\tau \in RT_k(K)$, and assume that*

(i) $\int_F \tau \cdot \mathbf{n}_K\, \psi = 0 \quad \forall\, \psi \in \mathbb{P}_k(F), \quad \forall F$ *face/side of K, when $k \geq 0$;*

(ii) $\int_K \tau \cdot \psi = 0 \quad \forall\, \psi \in [\mathbb{P}_{k-1}(K)]^n$, *when $k \geq 1$.*

Then $\tau \equiv 0$ in K.

Proof. We begin by observing that the number of degrees of freedom arising from (i) and (ii) reduces, respectively, to

$$(n+1)\,d_k = \frac{(n+1)(n+k-1)!}{(n-1)!\,k!}$$

and

$$n \dim \mathbb{P}_{k-1}(K) = n \binom{n+k-1}{k-1} = k\frac{(n+k-1)!}{(n-1)!\,k!},$$

and the sum of these quantities coincides with $\dim RT_k(K) = \dfrac{(n+k+1)(n+k-1)!}{(n-1)!k!}$. Hence, we remark in advance that the result provided by the present theorem guarantees the unisolvency of $RT_k(K)$ with respect to those degrees of freedom.

We now proceed with the proof itself. Let $\tau \in RT_k(K)$ such that (i) and (ii) are satisfied. Since $\tau \cdot \mathbf{n}_K|_F \in \mathbb{P}_k(F)$ for each face/side F of K, it follows directly from (i) that $\tau \cdot \mathbf{n}_K \equiv 0$ on ∂K. Furthermore, given $\varphi \in \mathbb{P}_k(K)$, we have $\nabla\varphi \in [\mathbb{P}_{k-1}(K)]^n$, and therefore

$$\int_K \varphi \operatorname{div}\tau = \begin{cases} \varphi \displaystyle\int_K \operatorname{div}\tau = \varphi \int_{\partial K} \tau\cdot\mathbf{n}_K = 0 & \text{if } k = 0, \\ -\displaystyle\int_K \nabla\varphi\cdot\tau + \langle \tau\cdot\mathbf{n}_K, \varphi\rangle_{\partial K} = -\int_K \nabla\varphi\cdot\tau = 0 & \text{if } k \geq 1, \end{cases}$$

where the last equality of the case $k \geq 1$ is a consequence of (ii). Thus, since $\operatorname{div}\tau \in \mathbb{P}_k(K)$, the preceding equation shows that

$$\operatorname{div}\tau \equiv 0 \quad \text{in} \quad K.$$

Next, let $q_1, q_2, \cdots, q_n \in \mathbb{P}_k(K)$ and $q_0 \in \tilde{\mathbb{P}}_k(K)$ such that

$$\tau(x) = \mathbf{q}(x) + x q_0(x) \qquad \forall x \in K,$$

where $\mathbf{q} := (q_1, q_2, \cdots, q_n)^{\mathtt{t}} \in [\mathbb{P}_k(K)]^n$. Then, if $q_0(x) = \sum_{|\alpha|=k} a_\alpha x^\alpha \quad \forall x \in K$, we find that

$$\begin{aligned} 0 = \operatorname{div}\tau(x) &= \operatorname{div}\mathbf{q}(x) + \sum_{i=1}^n \frac{\partial}{\partial x_i}(x_i q_0(x)) \\ &= \operatorname{div}\mathbf{q}(x) + n q_0(x) + \sum_{i=1}^n x_i \frac{\partial}{\partial x_i} q_0(x) \\ &= \operatorname{div}\mathbf{q}(x) + (n+k) q_0(x) \qquad \forall x \in K, \end{aligned}$$

from which

$$q_0 = -\frac{1}{(n+k)} \operatorname{div}\mathbf{q} \in \mathbb{P}_{k-1}(K),$$

which implies necessarily that $q_0 \equiv 0$, and hence $\tau = \mathbf{q} \in [\mathbb{P}_k(K)]^n$.

For the rest of the proof we assume for simplicity that $n = 2$. The case $n=3$ follows analogously. Since $\operatorname{div}\tau \equiv 0$ in K, there must exist a polynomial $w \in \mathbb{P}_{k+1}(K)$ (unique except for a constant) such that

$$\tau = \operatorname{curl} w := \left(\frac{\partial w}{\partial x_2}, -\frac{\partial w}{\partial x_1}\right)^{\mathtt{t}}.$$

Hence, denoting by s the tangential vector on ∂K, we have $0 = \tau\cdot\mathbf{n}_K = \operatorname{curl} w\cdot\mathbf{n}_K = \dfrac{dw}{ds}$ on ∂K, which yields $w \equiv$ constant on ∂K, and thus, without loss of generality, we can assume that $w \equiv 0$ on ∂K. The ultimate goal is to show from here that

$w \equiv 0$ in K so that τ vanishes as well in K. In fact, if $k \le 1$ (that is, $k+1 \le 2$), then the unisolvency of $\mathbb{P}_1(K)$ (with respect to the vertices of K) and $\mathbb{P}_2(K)$ (with respect to the vertices of K and half points of ∂K) guarantee that $w \equiv 0$ in K. Then, if $k \ge 2$ (that is, $k+1 \ge 3$), there exists $\lambda \in \mathbb{P}_{k-2}(K)$ such that $w = \lambda\, b_K$ in K, where $b_K \in \mathbb{P}_3(K)$ is the bubble function in K. In this way, denoting $\lambda(x) := \sum_{|\alpha|\le k-2} a_\alpha x^\alpha \quad \forall x \in K$, we define

$$\psi(x) := \left(0, \sum_{|\alpha|\le k-2} \frac{1}{\alpha_1+1} a_\alpha x_1^{\alpha_1+1} x_2^{\alpha_2} \cdots x_n^{\alpha_n}\right)^{\mathrm{t}} =: (\psi_1, \psi_2)^{\mathrm{t}} \quad \forall x \in K,$$

which satisfies $\psi \in [\mathbb{P}_{k-1}(K)]^2$ and $\operatorname{rot}\psi := \dfrac{\partial \psi_2}{\partial x_1} - \dfrac{\partial \psi_1}{\partial x_2} = \lambda$ in K. Thus, employing again hypothesis (ii), we deduce that

$$0 = \int_K \tau\cdot\psi = \int_K \mathbf{curl}\, w\cdot\psi = \int_K w \operatorname{rot}\psi = \int_K \lambda^2\, b_K,$$

which implies that $\lambda \equiv 0$ in K, and hence $w \equiv 0$ in K.

□

3.4 Interpolation in $H(\mathrm{div};\Omega)$

3.4.1 Local and Global Interpolation Operators

Given the triangularization $\mathcal{T}_h$ of $\overline{\Omega}$ introduced in Sect. 3.1, and given an integer $k \ge 0$, we define the global Raviart–Thomas space as

$$H_h^k := \left\{\tau \in H(\mathrm{div};\Omega): \quad \tau|_K \in RT_k(K) \quad \forall K \in \mathcal{T}_h\right\}.$$

Concerning this definition, it is important to recall from Theorem 3.2 that if τ is a vector function in $[L^2(\Omega)]^n$ such that $\tau|_K \in [H^1(K)]^n \quad \forall K \in \mathcal{T}_h$, then the necessary and sufficient condition for τ to belong to $H(\mathrm{div};\Omega)$ is that there holds $\tau\cdot\mathbf{n}_{K_i} + \tau\cdot\mathbf{n}_{K_j} = 0$ in $L^2(F) \quad \forall K_i, K_j \in \mathcal{T}_h$ adjacent with the common face/side F. In particular, if $\tau|_K \in RT_k(K) \quad \forall K \in \mathcal{T}_h$, then, according to part (ii) of Lemma 3.6, such a condition becomes

$$\int_F (\tau\cdot\mathbf{n}_{K_i} + \tau\cdot\mathbf{n}_{K_j})\, \psi_{l,F} = 0 \qquad \forall l \in \{1,2,\cdots,d_k\}, \tag{3.4}$$

where $\{\psi_{1,F}, \psi_{2,F}, \cdots, \psi_{d_{k,F}}\}$ is a basis of $\mathbb{P}_k(F)$. Equivalently, if we fix a normal vector to F such as $\mathbf{n}_F := \mathbf{n}_{K_i}$ or $\mathbf{n}_F := \mathbf{n}_{K_j}$, and note that $\mathbf{n}_{K_i} = -\mathbf{n}_{K_j}$, then (3.4) is rewritten as

$$\int_F \tau|_{K_i}\cdot\mathbf{n}_F\, \psi_{l,F} = \int_F \tau|_{K_j}\cdot\mathbf{n}_F\, \psi_{l,F} \qquad \forall l \in \{1,2,\cdots,d_k\}.$$

Hence, given $\tau \in H(\text{div};\Omega) \cap Z$ (cf. Theorem 3.2), that is, $\tau \in H(\text{div};\Omega)$ such that $\tau|_K \in [H^1(K)]^n \quad \forall K \in \mathscr{T}_h$, we define the *F-moments* for $k \geq 0$ as the values

$$\int_F \tau \cdot \mathbf{n}_F \, \psi_{l,F} \quad \forall l \in \{1,2,\cdots,d_k\}, \quad \forall \text{ face/side } F \text{ of } \mathscr{T}_h .$$

All the *F-moments* of $\mathscr{T}_h$ are denoted henceforth $m_i(\tau), \quad \forall i \in \{1,2,\cdots,N_1\}$, where $N_1 =$ the number of faces/sides of $\mathscr{T}_h$ times d_k. Next, we also define the *K-moments* for $k \geq 1$ as the values

$$\int_K \tau \cdot \psi_{j,K} \quad \forall j \in \{1,2,\cdots,r_k\}, \quad \forall K \in \mathscr{T}_h ,$$

where $r_k := \dim [\mathbb{P}_{k-1}(K)]^n = \dfrac{k\,(n+k-1)!}{(n-1)!\,k!}$ and $\{\psi_{1,K}, \psi_{2,K}, \cdots, \psi_{r_k,K}\}$ is a basis of $[\mathbb{P}_{k-1}(K)]^n$. All the *K-moments* of $\mathscr{T}_h$ are denoted $m_i(\tau)$, $\forall i \in \{N_1+1, N_1+2,\cdots,N\}$, where $N - N_1 =$ the number of polyhedra of $\mathscr{T}_h$ times r_k.

Now, given $j \in \{1,2,\cdots,N\}$, we let φ_j be the unique function in H_h^k such that

$$m_i(\varphi_j) = \delta_{ij} \qquad \forall i \in \{1,2,\cdots N\},$$

and introduce the global interpolation operator $\Pi_h^k : H(\text{div};\Omega) \cap Z \to H_h^k$ as

$$\Pi_h^k(\tau) := \sum_{j=1}^{N} m_j(\tau)\,\varphi_j \qquad \forall \tau \in H(\text{div};\Omega) \cap Z. \tag{3.5}$$

Equivalently, $\Pi_h^k(\tau)$ is the unique function in H_h^k such that

$$m_i(\Pi_h^k(\tau)) = m_i(\tau) \qquad \forall i \in \{1,2,\cdots,N\}.$$

Then for each $K \in \mathscr{T}_h$ we let $m_{i,K}(\tau)$, $i \in \{1,2,\cdots,N_K\}$, be the corresponding local moments, that is, the F-moments of the faces/sides F of K and the K-moments of K. Since the number of faces/sides of K is $n+1$, we have that $N_K = (n+1)d_k + r_k$. Then we define the local interpolation operator $\Pi_K^k : [H^1(K)]^n \to RT_k(K)$ as

$$\Pi_K^k(\tau) := \sum_{j=1}^{N_K} m_{j,K}(\tau)\,\varphi_{j,K} \quad \forall \tau \in [H^1(K)]^n, \tag{3.6}$$

where, given $j \in \{1,2,\cdots,N_K\}$, $\varphi_{j,K}$ is the unique function in $RT_k(K)$ such that

$$m_{i,K}(\varphi_{j,K}) = \delta_{i,j} \quad \forall i \in \{1,2,\cdots,N_K\}.$$

Note that there holds $\Pi_h^k(\tau)|_K = \Pi_K^k(\tau|_K) \quad \forall \tau \in H(\text{div};\Omega) \cap Z$.

The following lemma relates the divergences of the local and global interpolation operators in terms of the orthogonal projectors

$$\mathscr{P}_K^k : L^2(K) \to \mathbb{P}_k(K) \quad \text{and} \quad \mathscr{P}_h^k : L^2(\Omega) \to Y_h^k ,$$

where

$$Y_h^k := \left\{ v \in L^2(\Omega): \quad v|_K \in \mathbb{P}_k(K) \quad \forall K \in \mathscr{T}_h \right\}.$$

Lemma 3.7. *There holds*

$$\mathrm{div}\,(\Pi_K^k(\tau)) = \mathscr{P}_K^k(\mathrm{div}\,\tau) \qquad \forall \tau \in [H^1(K)]^n \tag{3.7}$$

and

$$\mathrm{div}\,(\Pi_h^k(\tau)) = \mathscr{P}_h^k(\mathrm{div}\,\tau) \qquad \forall \tau \in H(\mathrm{div};\Omega) \cap Z. \tag{3.8}$$

Proof. Given $\tau \in [H^1(K)]^n$, there clearly hold $\mathrm{div}\,(\Pi_K^k(\tau)) \in \mathbb{P}_k(K)$ and $m_{i,K}(\tau) = m_{i,K}(\Pi_K^k(\tau)) \quad \forall i \in \{1,2,\cdots,N_K\}$. Hence, for each $\psi \in \mathbb{P}_k(K)$ we obtain

$$\begin{aligned}\int_K \psi\,\mathrm{div}\,(\Pi_K^k(\tau)) &= -\int_K \nabla\psi \cdot \Pi_K^k(\tau) + \int_{\partial K} \psi\,\Pi_K^k(\tau)\cdot \mathbf{n}_K \\ &= -\int_K \nabla\psi \cdot \tau + \int_{\partial K} \psi\,\tau\cdot\mathbf{n}_K = \int_K \psi\,\mathrm{div}\,\tau,\end{aligned}$$

that is,

$$\int_K \psi\,\mathrm{div}\,(\Pi_K^k(\tau)) = \int_K \psi\,\mathrm{div}\,\tau \qquad \forall \psi \in \mathscr{P}_k(K), \tag{3.9}$$

which certainly proves (3.7). In addition, given $\tau \in H(\mathrm{div};\Omega) \cap Z$, we have that $\mathrm{div}\,(\Pi_h^k(\tau)) \in Y_h^k$, and hence for each $\psi \in Y_h^k$ we deduce, utilizing (3.9), that

$$\begin{aligned}\int_\Omega \psi\,\mathrm{div}\,(\Pi_h^k(\tau)) &= \sum_{K\in\mathscr{T}_h} \int_K \psi\,\mathrm{div}\,(\Pi_h^k(\tau)) \\ &= \sum_{K\in\mathscr{T}_h} \int_K \psi\,\mathrm{div}\,(\Pi_K^k(\tau)) \\ &= \sum_{K\in\mathscr{T}_h} \int_K \psi\,\mathrm{div}\,\tau = \int_\Omega \psi\,\mathrm{div}\,\tau,\end{aligned}$$

which implies (3.8) and completes the proof. □

Note that an alternative proof for (3.8) is obtained using that

$$\mathscr{P}_h^k(v)|_K = \mathscr{P}_K^k(v|_K) \qquad \forall v \in L^2(\Omega), \quad \forall K \in \mathscr{T}_h.$$

In fact, thanks to the preceding expression and identity (3.7), we obtain for each $K \in \mathscr{T}_h$

$$\mathscr{P}_h^k(\mathrm{div}\,\tau)|_K = \mathscr{P}_K^k(\mathrm{div}\,\tau) = \mathrm{div}\,(\Pi_K^k(\tau)) = \mathrm{div}\,(\Pi_h^k(\tau)|_K) = \mathrm{div}\,(\Pi_h^k(\tau))|_K,$$

which yields (3.8). The preceding identity is known in the literature as the COMMUTING PROPERTY, which is illustrated through the diagram

$$\begin{array}{ccc} H(\mathrm{div};\Omega)\cap Z & \stackrel{\mathrm{div}}{\longrightarrow} & L^2(\Omega) \\ \Pi_h^k \downarrow & \circlearrowleft & \downarrow \mathscr{P}_h^k \\ H_h^k & \stackrel{\mathrm{div}}{\longrightarrow} & Y_h^k \end{array}$$

We now aim to estimate the global interpolation error $\tau - \Pi_h^k(\tau)\ \forall \tau \in H(\mathrm{div};\Omega) \cap Z$ in terms of the corresponding local estimates for $\tau - \Pi_K^k(\tau)\ \forall \tau \in [H^1(K)]^n$. For this purpose, we require the concepts and results provided in the following two subsections.

3.4.2 Piola Transformation

Given $K \in \mathscr{T}_h$, $\tau \in [H^1(K)]^n$ and the affine mapping $T_K : \mathbb{R}^n \to \mathbb{R}^n$ defined by $T_K(\hat{x}) := B_K\hat{x} + b_K\ \forall \hat{x} \in \mathbb{R}^n$, with $B_K \in \mathbb{R}^{n\times n}$ invertible and $b_K \in \mathbb{R}^n$, such that $K = T_K(\hat{K})$, where $\hat{K}$ is the reference polyhedron (cf. Sect. 3.1), we introduce the PIOLA TRANSFORMATION:

$$\hat{\tau} := |\det B_K|\, B_K^{-1}\, \tau \circ T_K\,. \tag{3.10}$$

Note that the Gâteaux derivative $DT_K : \mathbb{R}^n \longrightarrow \mathscr{L}(\mathbb{R}^n, \mathbb{R}^n)$ defined by

$$DT_K(\hat{x})(\hat{y}) = \lim_{\varepsilon\to 0} \frac{T_K(\hat{x}+\varepsilon\hat{y}) - T_K(\hat{x})}{\varepsilon} \qquad \forall \hat{x},\hat{y} \in \mathbb{R}^n$$

reduces to

$$DT_K(\hat{x}) \equiv B_K \qquad \forall \hat{x} \in \mathbb{R}^n.$$

In turn, recall that if $F \in C^1(G)$, where G is an open set of $\mathbb{R}^n$, then

$$DF(x)(y) = \nabla F(x)\cdot y \qquad \forall x \in G, \quad \forall y \in \mathbb{R}^n,$$

and analogously for the case of a vector or tensor field F. In what follows we utilize the change of variable formula given by

$$\int_{\hat{K}} f(T_K(\hat{x}))\, d\hat{x} = \int_K |\det B_K|^{-1} f(x)\, dx \qquad \forall f \in C(K). \tag{3.11}$$

In addition, we make use of $\{\vec{e_1}, \vec{e_2}, \cdots, \vec{e_n}\}$, which denotes the canonical basis of $\mathbb{R}^n$.

Lemma 3.8. *Given $K \in \mathscr{T}_h$, there holds $\tau \in [H^1(K)]^n$ if and only if the Piola transformation $\hat{\tau} := |\det B_K|\, B_K^{-1}\, \tau \circ T_K \in [H^1(\hat{K})]^n$. In addition, given an integer $k \geq 0$, there holds $\tau \in RT_k(K)$ if and only if $\hat{\tau} \in RT_k(\hat{K})$.*

Proof. We observe first, thanks to (3.11), that

$$\int_{\hat K}\|\hat\tau(\hat x)\|^2\,d\hat x = \int_K |\det B_K|\,\|B_K^{-1}\,\tau(x)\|^2\,dx \le |\det B_K|\,\|B_K^{-1}\|^2\int_K \|\tau(x)\|^2\,dx,$$

which shows that $\hat\tau\in[L^2(\hat K)]^n$ if $\tau\in[L^2(K)]^n$.

On the other hand, using the chain rule we find for all $\hat x,\hat y\in\mathbb{R}^n$ that

$$\begin{aligned} D\hat\tau(\hat x)(\hat y) &= |\det B_K|\,B_K^{-1}D\tau(T_K(\hat x))(DT_K(\hat x)(\hat y))\\ &= |\det B_K|\,B_K^{-1}D\tau(T_K(\hat x))(B_K\hat y)\,. \end{aligned}$$

In particular, if $\hat y=\vec{e_j}$, then we obtain

$$\begin{aligned} \frac{\partial\hat\tau}{\partial\hat x_j}(\hat x) &= D\hat\tau(\hat x)(\vec{e_j}) = |\det B_K|\,B_K^{-1}D\tau(T_K(\hat x))(b_{K,j})\\ &= |\det B_K|\,B_K^{-1}\nabla\tau(T_K(\hat x))\,b_{K,j}\,, \end{aligned} \tag{3.12}$$

where $b_{K,j}$ is the jth column of B_K. Thus, employing again formula (3.11), we deduce that for each $j\in\{1,2,\cdots,n\}$ there holds

$$\begin{aligned} \int_{\hat K}\left\|\frac{\partial\hat\tau}{\partial\hat x_j}(\hat x)\right\|^2 d\hat x &= \int_K|\det B_K|\,\|B_K^{-1}\nabla\tau(x)\,b_{K,j}\|^2\,dx\\ &\le |\det B_K|\,\|B_K^{-1}\|^2\,\|b_{K,j}\|^2\int_K\|\nabla\tau(x)\|^2\,dx, \end{aligned}$$

which shows that $\nabla\hat\tau\in[L^2(\hat K)]^{n\times n}$ if $\nabla\tau\in[L^2(K)]^{n\times n}$. Actually, from the preceding analysis we conclude that $\hat\tau\in[H^1(\hat K)]^n$ whenever $\tau\in[H^1(K)]^n$. The converse is proved analogously.

Furthermore, let $\tau\in RT_k(K)$. It follows that there exist $\mathbf{p}\in[\mathbb{P}_k(K)]^n$ and $p_0\in\mathbb{P}_k(K)$ such that

$$\tau(x) = \mathbf{p}(x) + x\,p_0(x) \qquad \forall x\in K.$$

Hence,

$$\begin{aligned} \hat\tau(\hat x) &= |\det B_K|\,B_K^{-1}\tau(T_K(\hat x)) = |\det B_K|\,B_K^{-1}\tau(B_K\hat x+b_K)\\ &= |\det B_K|\,B_K^{-1}\Big\{\mathbf{p}(B_K\hat x+b_K) + p_0(B_K\hat x+b_K)(B_K\hat x+b_K)\Big\}\\ &= |\det B_K|\Big\{B_K^{-1}\,\mathbf{p}(B_K\hat x+b_K) + p_0(B_K\hat x+b_K)\,B_K^{-1}\,b_K\Big\}\\ &\quad + |\det B_K|\,\hat x\,p_0(B_K\hat x+b_K)\\ &= \hat{\mathbf{p}}(\hat x) + \hat x\,\hat p_0(\hat x) \qquad \forall\hat x\in\hat K, \end{aligned}$$

where $\hat{\mathbf{p}}\in[\mathbb{P}_k(\hat K)]^n$ and $\hat p_0\in\mathbb{P}_k(\hat K)$, which proves that $\hat\tau\in RT_k(\hat K)$. The converse follows directly from the inverse Piola transformation

$$\tau = |\det B_K|^{-1}\,B_K\,\hat\tau\circ T_K^{-1}.$$

□

It is important to remark that the second part of the preceding lemma would not hold if, instead of (3.10), one defined $\hat{\tau} := \tau \circ T_K$ (as for the Lagrange finite elements) (e.g., [14, 20]).

The following lemma establishes important identities involving the Piola transformation and the usual affine mapping for Lagrange elements.

Lemma 3.9. *Let $\tau \in [H^1(K)]^n$ and $\psi \in H^1(K)$, and let us define*

$$\hat{\tau} := |\det B_K|\, B_K^{-1}\, \tau \circ T_K \quad \text{and} \quad \hat{\psi} := \psi \circ T_K\,.$$

Then

(a) $\displaystyle\int_{\hat{K}} \hat{\tau} \cdot \nabla \hat{\psi} = \int_K \tau \cdot \nabla \psi$,

(b) $\displaystyle\int_{\hat{K}} \hat{\psi}\, \mathrm{div}\, \hat{\tau} = \int_K \psi\, \mathrm{div}\, \tau$,

(c) $\displaystyle\int_{\partial \hat{K}} \hat{\psi}\, \hat{\tau} \cdot \mathbf{n}_{\hat{K}} = \int_{\partial K} \psi\, \tau \cdot \mathbf{n}_K$.

Proof. Note first that, thanks to the chain rule,

$$D\hat{\psi}(\hat{x})(\hat{y}) = D\psi(T_K(\hat{x}))(B_K \hat{y}),$$

which yields, in particular,

$$\frac{\partial \hat{\psi}}{\partial \hat{x}_j}(\hat{x}) := D\hat{\psi}(\hat{x})(\vec{e_j}) = D\psi(T_K(\hat{x}))(b_{K,j}),$$

where $b_{K,j}$ is the jth column of B_K. It follows that

$$\frac{\partial \hat{\psi}}{\partial \hat{x}_j}(\hat{x}) = \nabla \psi(T_K(\hat{x})) \cdot b_{K,j} \qquad \forall j \in \{1, 2, \cdots, n\},$$

that is,

$$\frac{\partial \hat{\psi}}{\partial \hat{x}_j}(\hat{x}) = b_{K,j}^{\mathrm{t}}\, \nabla \psi(T_K(\hat{x})) \qquad \forall j \in \{1, 2, \cdots, n\},$$

whence

$$\nabla \hat{\psi}(\hat{x}) = B_K^{\mathrm{t}}\, \nabla \psi(T_K(\hat{x})) \qquad \forall \hat{x} \in \hat{K}.$$

Thus, employing the change of variable formula (3.11) we obtain

$$\begin{aligned}
\int_{\hat{K}} \hat{\tau} \cdot \nabla \hat{\psi} &= \int_{\hat{K}} |\det B_K|\, B_K^{-1}\, \tau \circ T_K \cdot B_K^{\mathrm{t}} \nabla \psi \circ T_K \\
&= \int_K B_K^{-1}\, \tau \cdot B_K^{\mathrm{t}} \nabla \psi = \int_K (\nabla \psi)^{\mathrm{t}}\, B_K B_K^{-1}\, \tau \\
&= \int_K (\nabla \psi)^{\mathrm{t}}\, \tau = \int_K \tau \cdot \nabla \psi,
\end{aligned}$$

which proves (a).

On the other hand, since [cf. (3.12)]

$$\frac{\partial \hat{\tau}}{\partial \hat{x}_j}(\hat{x}) = |\det B_K|\, B_K^{-1} \nabla \tau(T_K(\hat{x})) b_{K,j} \qquad \forall j \in \{1,2,\cdots,n\},$$

we obtain that

$$\nabla \hat{\tau}(\hat{x}) = |\det B_K|\, B_K^{-1} \nabla \tau(T_K(\hat{x}))\, B_K \qquad \forall \hat{x} \in \hat{K}.$$

Hence, using that $\mathrm{tr}(B^{-1}TB) = \mathrm{tr}(T)$, we deduce that

$$\begin{aligned} \mathrm{div}\, \hat{\tau}(\hat{x}) &= \mathrm{tr}\, \nabla \hat{\tau}(\hat{x}) = |\det B_K|\, \mathrm{tr}\, \nabla \tau(T_K(\hat{x})) \\ &= |\det B_K|\, \mathrm{div}\, \tau(T_K(\hat{x})) \qquad \forall \hat{x} \in \hat{K}, \end{aligned}$$

which implies

$$\mathrm{div}\, \hat{\tau} = |\det B_K|\, \mathrm{div}\, \tau \circ T_K \quad \text{in} \quad \hat{K}. \tag{3.13}$$

In this way, applying again (3.11), we find that

$$\int_{\hat{K}} \hat{\psi}\, \mathrm{div}\, \hat{\tau} = \int_{\hat{K}} \psi \circ T_K\, |\det B_K|\, \mathrm{div}\, \tau \circ T_K = \int_K \psi\, \mathrm{div}\, \tau,$$

which proves (b).

Finally, integrating by parts and utilizing (a) and (b) we conclude that

$$\begin{aligned} \int_{\partial \hat{K}} \hat{\psi}\, \hat{\tau} \cdot \mathbf{n}_{\hat{K}} &= \int_{\hat{K}} \hat{\tau} \cdot \nabla \hat{\psi} + \int_{\hat{K}} \hat{\psi}\, \mathrm{div}\, \hat{\tau} \\ &= \int_K \tau \cdot \nabla \psi + \int_K \psi\, \mathrm{div}\, \tau = \int_{\partial K} \psi\, \tau \cdot \mathbf{n}_K, \end{aligned}$$

which proves (c) and completes the proof.

□

The following result is a consequence of the identity (c) from the preceding lemma and the fact that $H^{1/2}(\partial K)$ is dense in $L^2(\partial K)$. We leave its proof as an exercise for the reader.

Lemma 3.10. *Let $\tau \in [H^1(K)]^n$ and $\varphi \in L^2(\partial K)$. Then*

$$\int_{\partial \hat{K}} \hat{\varphi}\, \hat{\tau} \cdot \mathbf{n}_{\hat{K}} = \int_{\partial K} \varphi\, \tau \cdot \mathbf{n}_K.$$

The relationship between the local interpolants on $K \in \mathscr{T}_h$ and the reference element $\hat{K}$ is established by the following lemma.

Lemma 3.11. *Given $K \in \mathscr{T}_h$ and $\tau \in [H^1(K)]^n$, there holds*

$$\Pi_{\hat{K}}^k(\hat{\tau}) = \widehat{\Pi_K^k(\tau)} := |\det B_K|\, B_K^{-1}\, \Pi_K^k(\tau) \circ T_K.$$

Proof. It suffices to show that the $\hat{F}$ and $\hat{K}$ moments of $\hat{\tau}$ and $\widehat{\Pi_K^k(\tau)}$ coincide. In fact, given $\hat{\psi} \in [\mathbb{P}_{k-1}(\hat{K})]^n$, we let $\psi := \hat{\psi} \circ T_K^{-1} \in [\mathbb{P}_{k-1}(K)]^n$ and utilize (3.11) and the fact that $(B_K^{-1})^{\mathtt{t}}\,\psi \in [\mathbb{P}_{k-1}(K)]^n$ to deduce that

$$\begin{aligned}
\int_{\hat{K}} \widehat{\Pi_K^k(\tau)} \cdot \hat{\psi} &= \int_{\hat{K}} |\det B_K|\, B_K^{-1}\, \Pi_K^k(\tau) \circ T_K \cdot \psi \circ T_K \\
&= \int_K B_K^{-1}\, \Pi_K^k(\tau) \cdot \psi = \int_K \Pi_K^k(\tau) \cdot (B_K^{-1})^{\mathtt{t}}\, \psi \\
&= \int_K \tau \cdot (B_K^{-1})^{\mathtt{t}}\, \psi = \int_K B_K^{-1}\, \tau \cdot \psi \\
&= \int_{\hat{K}} |\det B_K|\, B_K^{-1}\, \tau \circ T_K \cdot (\psi \circ T_K) = \int_{\hat{K}} \hat{\tau} \cdot \hat{\psi},
\end{aligned}$$

which proves that the $\hat{K}$-moments of $\widehat{\Pi_K^k(\tau)}$ and $\hat{\tau}$ coincide. Then, given a face/side $\hat{F}$ of $\partial\hat{K}$ and $\hat{\psi} \in \mathbb{P}_k(\hat{F})$, we extend $\hat{\psi}$ by zero on $\partial\hat{K}\backslash\hat{F}$ so that the resulting $\hat{\psi}$ belongs to $L^2(\partial\hat{K})$. It follows, using Lemma 3.10 and the fact that the F-moments of τ and $\Pi_K^k(\tau)$ are the same, and denoting $F = T_K(\hat{F})$, that

$$\begin{aligned}
\int_{\hat{F}} \widehat{\Pi_K^k(\tau)} \cdot \mathbf{n}_{\hat{K}}\, \hat{\psi} &= \int_{\partial\hat{K}} \widehat{\Pi_K^k(\tau)} \cdot \mathbf{n}_{\hat{K}}\, \hat{\psi} \\
&= \int_{\partial K} \Pi_K^k(\tau) \cdot \mathbf{n}_K\, \psi = \int_F \Pi_K^k(\tau) \cdot \mathbf{n}_K\, \psi \\
&= \int_F \tau \cdot \mathbf{n}_K\, \psi = \int_{\partial K} \tau \cdot \mathbf{n}_K\, \psi \\
&= \int_{\partial\hat{K}} \hat{\tau} \cdot \mathbf{n}_{\hat{K}}\, \hat{\psi} = \int_{\hat{F}} \hat{\tau} \cdot \mathbf{n}_{\hat{K}}\, \hat{\psi},
\end{aligned}$$

which shows that the $\hat{F}$-moments of $\hat{\tau}$ and $\widehat{\Pi_K^k(\tau)}$ coincide, thus completing the proof.

□

3.4.3 Deny–Lions, Bramble–Hilbert, and Related Results

In this section we recall some fundamental results on the interpolation theory of the usual Sobolev spaces, which will also be employed for the present analysis in $H(\mathrm{div};\Omega)$.

Given an integer $k \geq 0$ and a compact and connected set S in $\mathbb{R}^n$ with a Lipschitz-continuous boundary, we are interested in the quotient space $H^{k+1}(S)/\mathbb{P}_k(S)$ given by

$$H^{k+1}(S)/\mathbb{P}_k(S) := \Big\{ [v] : \quad v \in H^{k+1}(S) \Big\},$$

where

$$[v] := \left\{ w \in H^{k+1}(S) : \quad v - w \in \mathbb{P}_k(S) \right\}.$$

It is well known that $H^{k+1}(S)/\mathbb{P}_k(S)$, endowed with the norm

$$\|[v]\|_{k+1,k,S} := \inf_{w\in[v]} \|w\|_{k+1,S} = \inf_{p\in\mathbb{P}_k(S)} \|v+p\|_{k+1,S} := \mathrm{dist}(v,\mathbb{P}_k(S))$$

for all $[v] \in H^{k+1}(S)/\mathbb{P}_k(S)$, is a Banach space. Note that $\|\cdot\|_{k+1,k,S}$ is well defined since whenever $v-w \in \mathbb{P}_k(S)$ there clearly holds $\mathrm{dist}(v,\mathbb{P}_k(S)) = \mathrm{dist}(w,\mathbb{P}_k(S))$. In addition, the mapping

$$\begin{aligned} |\cdot|_{k+1,k,S} \,:\, H^{k+1}(S)/\mathbb{P}_k(S) &\longrightarrow \mathbb{R}^+ \\ [v] &\longrightarrow |[v]|_{k+1,k,S} := |v|_{k+1,S} \end{aligned}$$

is also well defined since the fact that $v-w \in \mathbb{P}_k(S)$ yields $\partial^\alpha(v-w) = 0\ \forall\alpha\in\mathbb{N}_0^n$, with $|\alpha| = k+1$, and therefore

$$|w|_{k+1,S} = |v+(w-v)|_{k+1,S} = |v|_{k+1,S}.$$

Then, for each $p \in \mathbb{P}_k(S)$ and for each $v \in H^{k+1}(S)$ we have

$$\|v+p\|_{k+1,S}^2 = \|v+p\|_{k,S}^2 + |v+p|_{k+1,S}^2 = \|v+p\|_{k,S}^2 + |v|_{k+1,S}^2 \geq |v|_{k+1,S}^2,$$

from which

$$|[v]|_{k+1,k,S} := |v|_{k+1,S} \leq \mathrm{dist}(v,\mathbb{P}_k(S)) =: \|[v]\|_{k+1,k,S}.$$

The following result provides the converse inequality (except for a constant), thanks to which the seminorm $|\cdot|_{k+1,k,S}$ and the norm $\|\cdot\|_{k+1,k,S}$ become equivalent in $H^{k+1}(S)/\mathbb{P}_k(S)$.

Theorem 3.4 (Deny–Lions Lemma). *There exists $C > 0$, depending only on S, such that*

$$\|[v]\|_{k+1,k,S} \leq C\,|[v]|_{k+1,k,S} \qquad \forall v \in H^{k+1}(S).$$

Proof. Let $N := \dim\mathbb{P}_k(S) = \binom{n+k}{k}$, and let $\{f_1, f_2, \cdots, f_N\}$ be a basis of the dual of $\mathbb{P}_k(S)'$. Then it is clear that

$$\mathbb{P}_k(S) \cap {}^\circ\left\{f_1, f_2, \cdots, f_N\right\} = \left\{0\right\},$$

where the superscript $^\circ$ denotes the annihilator of the given set (cf. [54]). Hence, since $\mathbb{P}_k(S)$ is a subspace of $H^{k+1}(S)$, the Hahn–Banach theorem guarantees the existence of $\{F_1, F_2, \cdots, F_N\} \subseteq H^{k+1}(S)'$ such that

$$\|F_i\|_{H^{k+1}(S)'} = \|f_i\|_{\mathbb{P}_k(S)'} \quad \text{and} \quad F_i|_{\mathbb{P}_k(S)} = f_i \qquad \forall i \in \{1, 2, \cdots, N\}.$$

It follows that

$$\mathbb{P}_k(S) \cap {}^{\circ}\Big\{F_1, F_2, \cdots, F_N\Big\} = \Big\{0\Big\},$$

and therefore, the generalized Poincaré inequality (cf. [46, Theorem 5.11.2]) implies the existence of a constant $C = C(S) > 0$ such that

$$\|v\|_{k+1,S} \le C \left\{ |v|_{k+1,S}^2 + \sum_{i=1}^{N} |F_i(v)|^2 \right\}^{1/2} \qquad \forall v \in H^{k+1}(S).$$

Next, given scalars $\alpha_1, \alpha_2, \cdots, \alpha_N \in \mathbb{R}$, there exists a unique $q \in \mathbb{P}_k(S)$ such that $f_i(q) = \alpha_i \quad \forall i \in \{1,2,\cdots,N\}$. Indeed, it suffices to define $q := \sum_{j=1}^{N} \alpha_j q_j$, where $q_j \in \mathbb{P}_k(S)$ is such that $f_i(q_j) = \delta_{ij} \quad \forall i,j \in \{1,2,\cdots,N\}$. In this way, given $v \in H^{k+1}(S)$, there exists a unique $q_v \in \mathbb{P}_k(S)$ such that $f_i(q_v) = -F_i(v) \quad \forall i \in \{1,2,\cdots,N\}$. Consequently,

$$\begin{aligned} \|[v]\|_{k+1,k,S} &:= \inf_{p \in \mathbb{P}_k(S)} \|v+p\|_{k+1,S} \le \|v+q_v\|_{k+1,S} \\ &\le C \left\{ |v+q_v|_{k+1,S}^2 + \sum_{i=1}^{N} |F_i(v+q_v)|^2 \right\}^{1/2} \\ &= C|v+q_v|_{k+1,S} = C|v|_{k+1,S} = C|[v]|_{k+1,k,S}, \end{aligned}$$

which completes the proof. □

Note that the preceding proof could also be done in a bit simpler way by employing the Riesz representation theorem instead of the Hahn–Banach theorem. We leave this as an exercise for the reader.

The following result establishes a simple but, at the same time, fundamental boundedness property for operators that are defined in Sobolev spaces and that preserve polynomials.

Theorem 3.5 (Bramble–Hilbert Lemma). *Let m and k be nonnegative integers such that $0 \le m \le k+1$, and let $\Pi \in \mathscr{L}(H^{k+1}(S), H^m(S))$ such that $\Pi(p) = p \quad \forall p \in \mathbb{P}_k(S)$. Then there exists $C := C(\Pi, S) > 0$ such that*

$$\|v - \Pi(v)\|_{m,S} \le C|v|_{k+1,S} \qquad \forall v \in H^{k+1}(S).$$

Proof. Given $v \in H^{k+1}(S)$ and $p \in \mathbb{P}_k(S)$, we have

$$v - \Pi(v) = (v+p) - \Pi(v+p) = (I - \Pi)(v+p),$$

which, using that $I \in \mathscr{L}(H^{k+1}(S), H^m(S))$ since $0 \le m \le k+1$, implies

$$\|v - \Pi(v)\|_{m,S} \le \|I - \Pi\| \, \|v+p\|_{k+1,S} \qquad \forall p \in \mathbb{P}_k(S),$$

and therefore

$$\|v-\Pi(v)\|_{m,S} \leq \|I-\Pi\| \inf_{p\in\mathbb{P}_k(S)} \|v+p\|_{k+1,S} = \|I-\Pi\| \, \|[v]\|_{k+1,k,S}.$$

This inequality and the Deny–Lions lemma (cf. Theorem 3.4) complete the proof. □

On the other hand, the following two lemmas provide equivalence relationships between Sobolev spaces defined on affine-equivalent and Piola-equivalent domains.

Lemma 3.12. *Let S and $\hat{S}$ be compact and connected sets of $\mathbb{R}^n$ with Lipschitz-continuous boundaries, and let $F:\mathbb{R}^n \longrightarrow \mathbb{R}^n$ be the affine mapping given by $F(\hat{x}) = B\hat{x} + b \;\; \forall \hat{x}\in\mathbb{R}^n$, with $B\in\mathbb{R}^{n\times n}$ invertible and $b\in\mathbb{R}^n$, such that $S = F(\hat{S})$. Then let m be a nonnegative integer, and let $v\in H^m(S)$. Then $\hat{v} := v\circ F \in H^m(\hat{S})$, and there exists $C := C(m,n) > 0$ such that*

$$|\hat{v}|_{m,\hat{S}} \leq \hat{C}\,\|B\|^m\,|\det B|^{-1/2}\,|v|_{m,S}. \tag{3.14}$$

Conversely, if $\hat{v}\in H^m(\hat{S})$ and we let $v = \hat{v}\circ F^{-1}$, then $v\in H^m(S)$, and there exists $\hat{C} := \hat{C}(m,n) > 0$ such that

$$|v|_{m,S} \leq \hat{C}\,\|B^{-1}\|^m\,|\det B|^{1/2}\,|\hat{v}|_{m,\hat{S}}. \tag{3.15}$$

Proof. We use that $C^m(\overline{S})$ is dense in $H^m(S)$. Then, given $v\in C^m(\overline{S})$ and a multi-index α with $|\alpha| = m$, we have $\hat{v} := v\circ F \in C^m(\overline{\hat{S}})$ and

$$\partial^\alpha \hat{v}(\hat{x}) = D^m \hat{v}(\hat{x})(e_{\beta_1}, e_{\beta_2}, \cdots, e_{\beta_m}) \qquad \forall \hat{x}\in\hat{S},$$

where $\{e_{\beta_1}, e_{\beta_2}, \cdots, e_{\beta_m}\} \subseteq \{\vec{e_1}, \vec{e_2}, \cdots, \vec{e_n}\}$, the canonical basis of $\mathbb{R}^n$. It follows that

$$|\partial^\alpha \hat{v}(\hat{x})| \leq \sup_{\substack{\|\xi_i\|\leq 1\\ i\in\{1,2,\cdots,m\}}} |D^m \hat{v}(\hat{x})(\xi_1,\xi_2,\cdots,\xi_m)| =: \|D^m \hat{v}(\hat{x})\|,$$

and hence

$$\begin{aligned} |\hat{v}|^2_{m,\hat{S}} &= \int_{\hat{S}} \sum_{|\alpha|=m} |\partial^\alpha \hat{v}(\hat{x})|^2\, d\hat{x} \\ &\leq \sum_{|\alpha|=m} \int_{\hat{S}} \|D^m \hat{v}(\hat{x})\|^2\, d\hat{x} \\ &= C_1(m,n) \int_{\hat{S}} \|D^m \hat{v}(\hat{x})\|^2\, d\hat{x}, \end{aligned} \tag{3.16}$$

where $C_1(m,n) := \mathrm{card}\{\alpha : \;\; |\alpha| = m\}$. Now, utilizing the chain rule and the fact that $DF(\hat{x}) \equiv B \;\forall \hat{x}\in\mathbb{R}^n$, we deduce that

$$D^m \hat{v}(\hat{x})(\xi_1,\xi_2,\cdots,\xi_m) = D^m v(F(\hat{x}))(B\xi_1, B\xi_2,\cdots,B\xi_m)$$

for all $(\xi_1, \xi_2, \cdots, \xi_m) \in \mathbb{R}^n \times \cdots \times \mathbb{R}^n$, from which, denoting $x = F(\hat{x})$, we obtain

$$\|D^m \hat{v}(\hat{x})\| := \sup_{\substack{\|\xi_i\| \le 1 \\ i \in \{1,2,\cdots,m\}}} |D^m v(x)(B\xi_1, B\xi_2, \cdots, B\xi_m)|$$

$$= \|B\|^m \sup_{\substack{\|\xi_i\| \le 1 \\ i \in \{1,2,\cdots,m\}}} \left| D^m v(x) \left(\frac{B\xi_1}{\|B\|}, \frac{B\xi_2}{\|B\|}, \cdots, \frac{B\xi_m}{\|B\|} \right) \right|$$

$$\le \|B\|^m \sup_{\substack{\|\lambda_i\| \le 1 \\ i \in \{1,2,\cdots,m\}}} |D^m v(x)(\lambda_1, \lambda_2, \cdots, \lambda_m)| = \|B\|^m \|D^m v(x)\|.$$

In this way, employing also (3.11), we find from (3.16) that

$$|\hat{v}|^2_{m,\hat{S}} \le C_1(m,n)\, \|B\|^{2m} \int_{\hat{S}} \|D^m v(F(\hat{x}))\|^2 \, d\hat{x}$$

$$= C_1(m,n)\, \|B\|^{2m} \, |\det \mathrm{B}|^{-1} \int_S \|D^m v(x)\|^2 \, dx,$$

and since

$$\|D^m v(x)\| \le C_2(n) \max_{|\alpha|=m} |\partial^\alpha v(x)| \le C_2(n) \sum_{|\alpha|=m} |\partial^\alpha v(x)|,$$

we obtain

$$|\hat{v}|^2_{m,\hat{S}} \le C_3(m,n)\, \|B\|^{2m} \, |\det \mathrm{B}|^{-1} \, |v|^2_{m,S},$$

which proves (3.14) for $v \in C^m(\overline{S})$. Analogously, exchanging the roles of S and $\hat{S}$ and using F^{-1} instead of F, we have (3.15) for all $\hat{v} \in C^m(\overline{\hat{S}})$.

Similarly, for each $p \le m$ there hold

$$|\hat{v}|_{p,\hat{S}} \le C(p,n)\, \|B\|^p \, |\det B|^{-1/2} \, |v|_{p,S}$$

and

$$|v|_{p,S} \le C(p,n)\, \|B^{-1}\|^p \, |\det B|^{1/2} |\hat{v}|_{p,\hat{S}}$$

for all $v \in C^p(\overline{S})$, with $\hat{v} := v \circ F \in C^p(\overline{\hat{S}})$, which implies the existence of constants $C_i = C_i(m,n,B)$, $i \in \{1,2\}$, such that

$$C_1 \, \|\hat{v}\|_{m,\hat{S}} \le \|v\|_{m,S} \le C_2 \, \|\hat{v}\|_{m,\hat{S}} \qquad \forall v \in C^m(\overline{S}). \tag{3.17}$$

Now, given $v \in H^m(S)$, we consider a sequence $\{v_j\}_{j \in \mathbb{N}} \subseteq C^m(\overline{S})$ such that $\|v_j - v\|_{m,S} \stackrel{j\to\infty}{\longrightarrow} 0$. Thus, we obtain from (3.17) that

$$\|\hat{v}_j - \hat{v}_k\|_{m,\hat{S}} \le C_1^{-1} \, \|v_j - v_k\|_{m,S},$$

from which we deduce the existence of $\hat{v} \in H^m(\hat{S})$ such that $\|\hat{v}_j - \hat{v}\|_{m,\hat{S}} \stackrel{j\to\infty}{\longrightarrow} 0$. Moreover, it is easy to see that this limit $\hat{v}$ is independent of the chosen sequence, and hence we can define the operator

$$\begin{aligned} H^m(S) &\longrightarrow H^m(\hat{S}) \\ v &\longrightarrow \hat{v} := \text{“}v\circ F\text{”}. \end{aligned}$$

Finally, taking limit in the inequality (3.14) with $v = v_j$, that is,

$$|\hat{v}_j|_{m,\hat{S}} \leq \hat{C}\,\|B\|^m\,|\det B|^{-1/2}\,|v_j|_{m,S},$$

we arrive at

$$|\hat{v}|_{m,\hat{S}} \leq \hat{C}\,\|B\|^m\,|\det B|^{-1/2}\,|v|_{m,S},$$

which proves (3.14) $\forall v \in H^m(S)$. Analogously we prove (3.15) $\forall \hat{v} \in H^m(\hat{S})$.

□

Lemma 3.13. *Let S and $\hat{S}$ be compact and connected sets of $\mathbb{R}^n$ with Lipschitz-continuous boundaries, and let $F : \mathbb{R}^n \longrightarrow \mathbb{R}^n$ be the affine mapping given by $F(\hat{x}) = B\hat{x} + b \;\; \forall \hat{x} \in \mathbb{R}^n$, with $B \in \mathbb{R}^{n\times n}$ invertible and $b \in \mathbb{R}^n$, such that $S = F(\hat{S})$. In turn, let m be a nonnegative integer, and let $\tau \in [H^m(S)]^n$. Then $\hat{\tau} := |\det B|\,B^{-1}\,\tau\circ F \in [H^m(\hat{S})]^n$, and there exists $C := C(m,n) > 0$ such that*

$$|\hat{\tau}|_{m,\hat{S}} \leq C\,\|B^{-1}\|\,\|B\|^m\,|\det B|^{1/2}\,|\tau|_{m,S}. \tag{3.18}$$

Conversely, if $\hat{\tau} \in [H^m(\hat{S})]^n$ and we let $\tau := |\det B|^{-1}\,B\,\hat{\tau}\circ F^{-1}$, then $\tau \in [H^m(S)]^n$, and there exists $\hat{C} := \hat{C}(m,n) > 0$ such that

$$|\tau|_{m,S} \leq \hat{C}\,\|B\|\,\|B^{-1}\|^m\,|\det B|^{-1/2}\,|\hat{\tau}|_{m,\hat{S}}. \tag{3.19}$$

Proof. Following as in the proof of the previous lemma, it suffices to prove (3.18) for $\tau \in [C^m(\overline{S})]^n$. In fact, note first that for each α there holds

$$\partial^\alpha\,\hat{\tau}(\hat{x}) = |\det B|\,B^{-1}\,\partial^\alpha(\tau\circ F)(\hat{x}) \qquad \forall \hat{x} \in (\hat{S}),$$

which yields

$$\|\partial^\alpha\,\hat{\tau}\|_{0,\hat{S}} \leq |\det B|\,\|B^{-1}\|\,\|\partial^\alpha(\tau\circ F)\|_{0,\hat{S}}.$$

Hence, summing up over all the multi-indexes α with $|\alpha| = m$ we obtain

$$|\hat{\tau}|_{m,\hat{S}} \leq |\det B|\,\|B^{-1}\|\,|\tau\circ F|_{m,\hat{S}},$$

and applying the estimate provided by Lemma 3.12 to each of the components of $\tau\circ F$ we deduce that

$$\begin{aligned} |\hat{\tau}|_{m,\hat{S}} &\leq |\det B|\,\|B^{-1}\|\,C\,\|B\|^m\,|\det B|^{-1/2}\,|\tau|_{m,S} \\ &= C\,\|B^{-1}\|\,\|B\|^m\,|\det B|^{1/2}\,|\tau|_{m,S}, \end{aligned}$$

which, as stated, proves (3.18) for $\tau \in [C^m(\overline{S})]^n$ and completes the proof.

□

Next, to complete the estimates provided by the two previous lemmas, we need to bound $|\det B|$, $\|B\|$ and $\|B^{-1}\|$ in terms of the geometric properties of S. More precisely, we have the following result (e.g., [14, 20, 50]).

Lemma 3.14. *Let S and $\hat{S}$ be compact and connected sets of $\mathbb{R}^n$ with Lipschitz-continuous boundaries, and let $F : \mathbb{R}^n \longrightarrow \mathbb{R}^n$ be the affine mapping given by $F(\hat{x}) = B\hat{x} + b \quad \forall \hat{x} \in \mathbb{R}^n$, with $B \in \mathbb{R}^{n\times n}$ invertible and $b \in \mathbb{R}^n$, such that $S = F(\hat{S})$. Next, let*

$$\begin{aligned} h_S &:= \textit{diameter of } S = \max_{x,y\in S} \|x-y\|, \\ \rho_S &:= \textit{diameter of largest sphere contained in } S, \\ \hat{h} &:= \textit{diameter of } \hat{S}, \textit{ and} \\ \hat{\rho} &:= \textit{diameter of largest sphere contained in } \hat{S}. \end{aligned}$$

Then

$$|\det B| = \frac{|S|}{|\hat{S}|}, \quad \|B\| \leq \frac{h_S}{\hat{\rho}} \quad \textit{and} \quad \|B^{-1}\| \leq \frac{\hat{h}}{\rho_S}.$$

Proof. The identity for $|\det B|$ follows from the application of the change of variable formula [cf. (3.11)]

$$\int_{\hat{S}} f(F(\hat{x}))\, d\hat{x} = \int_S |\det B|^{-1} f(x)\, dx$$

to the function $f \equiv 1$ in S. On the other hand, we clearly have

$$\|B\| = \sup_{\substack{x\in\mathbb{R}^n \\ x\neq 0}} \frac{\|Bx\|}{\|x\|} = \frac{1}{\hat{\rho}} \sup_{\substack{x\in\mathbb{R}^n \\ \|x\|=\hat{\rho}}} \|Bx\|.$$

Now, given $x \in \mathbb{R}^n$, with $\|x\| = \hat{\rho}$, there exist $\hat{y}, \hat{z} \in \hat{S}$ such that $x = \hat{y} - \hat{z}$, and hence $Bx = B\hat{y} - B\hat{z} = F(\hat{y}) - F(\hat{z})$, with $F(\hat{y}),\, F(\hat{z}) \in S$. Thus, we have that $\|Bx\| = \|F(\hat{y}) - F(\hat{z})\| \leq h_S$, and therefore $\|B\| \leq h_S/\hat{\rho}$. Similarly, the estimate for $\|B^{-1}\|$ is obtained from the preceding expressions exchanging the roles of S and $\hat{S}$ through the inverse affine mapping F^{-1}.

□

3.4.4 Interpolation Errors

In what follows, and in order to derive the estimates for the interpolation errors, we apply the results from Sect. 3.4.3 to the polyhedra K of the triangularization $\mathscr{T}_h$ of $\overline{\Omega}$. We begin by proving the boundedness of the local interpolation operators.

Lemma 3.15. *Let m and k be integers such that $k \geq 0$. Then*

$$\Pi_K^k \in \mathscr{L}([H^{k+1}(K)]^n, [H^m(K)]^n) \qquad \forall K \in \mathscr{T}_h$$

and

$$\Pi_{\hat{K}}^k \in \mathscr{L}([H^{k+1}(\hat{K})]^n, [H^m(\hat{K})]^n).$$

Proof. It suffices to demonstrate for $K \in \mathscr{T}_h$. Recall first from Sect. 3.4.1 [cf. (3.6)] that, given $\tau \in [H^{k+1}(K)]^n$, we have

$$\Pi_K^k(\tau) = \sum_{i=1}^{N_K} m_{i,K}(\tau)\, \varphi_{i,K},$$

where $\{\varphi_{1,K}, \varphi_{2,K}, \cdots, \varphi_{N_K,K}\}$ is the canonical basis of $RT_k(K)$, and the local moments $m_{i,K}(\tau)$, $i \in \{1,2,\cdots,N_K\}$, are defined as

$$m_{i,K}(\tau) = \begin{cases} \displaystyle\int_F \tau \cdot \mathbf{n}_F\, \psi_{l,F} & \text{if it is an } F\text{-moment}, \\ \displaystyle\int_K \tau \cdot \psi_{j,K} & \text{if it is a } K\text{-moment}, \end{cases}$$

with $\{\psi_{1,F}, \psi_{2,F}, \cdots, \psi_{d_k,F}\}$ the basis of $\mathbb{P}_k(F)$ when $k \geq 0$ and $\{\psi_{1,K}, \psi_{2,K}, \cdots, \psi_{r_k,K}\}$ the basis of $[\mathbb{P}_{k-1}(K)]^n$ when $k \geq 1$. It follows that Π_K^k is linear and that

$$\left\|\Pi_K^k(\tau)\right\|_{m,K} \leq \sum_{i=1}^{N_K} |m_{i,K}(\tau)|\, \|\varphi_{i,K}\|_{m,K}. \qquad (3.20)$$

In addition, utilizing the trace theorem in $[H^1(K)]^n$ with boundedness constant $c(K)$ (cf. Theorems 1.4 and 1.5) and the Cauchy–Schwarz inequality, we obtain

$$\begin{aligned} \left|\int_F \tau \cdot \mathbf{n}_F\, \psi_{l,F}\right| &\leq \|\psi_{l,F}\|_{0,F}\, \|\tau \cdot \mathbf{n}_F\|_{0,F} \leq \|\psi_{l,F}\|_{0,F}\, \|\tau\|_{0,\partial K} \\ &\leq c(K)\, \|\psi_{l,F}\|_{0,F}\, \|\tau\|_{1,K} \leq c(K)\, \|\psi_{l,F}\|_{0,F}\, \|\tau\|_{k+1,K}, \end{aligned}$$

and

$$\left|\int_K \tau \cdot \psi_{j,K}\right| \leq \|\psi_{j,K}\|_{0,K}\, \|\tau\|_{0,K} \leq \|\psi_{j,K}\|_{0,K}\, \|\tau\|_{k+1,K}.$$

From these two estimates and (3.20) we conclude that Π_K^k is bounded with $\|\Pi_K^k\|$ depending on $c(K)$, $\|\varphi_{i,K}\|_{m,K}$, $i \in \{1,2,\cdots,N_K\}$, $\|\psi_{l,F}\|_{0,F}$, $l \in \{1,2,\cdots,d_k\}$, and $\|\psi_{j,K}\|_{0,K}$, $j \in \{1,2,\cdots,r_k\}$.

□

We now establish the first error estimate.

Lemma 3.16 (Local Interpolation Error). *Let m and k be nonnegative integers such that $0 \leq m \leq k+1$. Then there exists $C := C(\hat{K}, \Pi_{\hat{K}}^k, k, m, n) > 0$ such that*

$$|\tau - \Pi_K^k(\tau)|_{m,K} \leq C\, \frac{h_K^{k+2}}{\rho_K^{m+1}}\, |\tau|_{k+1,K} \qquad \forall\, \tau \in [H^{k+1}(K)]^n. \qquad (3.21)$$

In addition, for each $\tau \in [H^1(K)]^n$ *with* $\operatorname{div}\tau \in H^{k+1}(K)$ *there holds*

$$|\operatorname{div}\tau - \operatorname{div}\Pi_K^k(\tau)|_{m,K} \leq C\,\frac{h_K^{k+1}}{\rho_K^m}\,|\operatorname{div}\tau|_{k+1,K}\,. \tag{3.22}$$

Proof. Let $\tau \in [H^{k+1}(K)]^n$. Employing the estimate (3.19) (cf. Lemma 3.13) and the fact that $\Pi_{\hat K}^k(\hat\tau) = \widehat{\Pi_K^k(\tau)}$ (cf. Lemma 3.11) we obtain

$$|\tau - \Pi_K^k(\tau)|_{m,K} \leq C\,\|B_K\|\,\|B_K^{-1}\|^m\,|\det B_K|^{-1/2}\,|\hat\tau - \Pi_{\hat K}^k(\hat\tau)|_{m,\hat K}\,. \tag{3.23}$$

Hence, since $\Pi_{\hat K}^k \in \mathscr{L}([H^{k+1}(\hat K)]^n, [H^m(\hat K)]^n)$ (cf. Lemma 3.15), $\Pi_{\hat K}^k(\hat{\mathbf{p}}) = \hat{\mathbf{p}}\ \forall \hat{\mathbf{p}} \in RT_k(\hat K)$, and $[\mathbb{P}_k(\hat K)]^n \subseteq RT_k(\hat K)$, the Bramble–Hilbert lemma implies that

$$|\hat\tau - \Pi_{\hat K}^k(\hat\tau)|_{m,\hat K} \leq C\,|\hat\tau|_{k+1,\hat K}\,. \tag{3.24}$$

Then, applying the estimate (3.18) (cf. Lemma 3.13), we obtain

$$|\hat\tau|_{k+1,\hat K} \leq C\,\|B_K^{-1}\|\,\|B_K\|^{k+1}\,|\det B_K|^{1/2}\,|\tau|_{k+1,K}\,. \tag{3.25}$$

Thus, inserting (3.25) into (3.24) and then the resulting bound into (3.23) we deduce that

$$|\tau - \Pi_K^k(\tau)|_{m,K} \leq C\,\|B_K\|^{k+2}\,\|B_K^{-1}\|^{m+1}\,|\tau|_{k+1,K}\,,$$

from which, using that $\|B_K^{-1}\| \leq \dfrac{\hat h}{\rho_K}$ and $\|B_K\| \leq \dfrac{h_K}{\hat\rho}$ (cf. Lemma 3.14), we arrive at (3.21).

On the other hand, let $\tau \in [H^1(K)]^n$, with $\operatorname{div}\tau \in H^{k+1}(K)$. Then, recalling from Sect. 3.4.2 [cf. (3.13)] that

$$\operatorname{div}\hat\tau = |\det B_K|\operatorname{div}\tau \circ T_K \qquad \forall\,\tau \in [H^1(K)]^n\,, \tag{3.26}$$

and then, utilizing also Lemma 3.11, we find that

$$\begin{aligned}
\widehat{\operatorname{div}\tau - \operatorname{div}\Pi_K^k(\tau)} &= \widehat{\operatorname{div}\tau} - \widehat{\operatorname{div}\Pi_K(\tau)}\\
&= |\det B_K|^{-1}\left\{\operatorname{div}\hat\tau - \operatorname{div}\widehat{\Pi_K^k(\tau)}\right\}\\
&= |\det B_K|^{-1}\left\{\operatorname{div}\hat\tau - \operatorname{div}\Pi_{\hat K}^k(\hat\tau)\right\}.
\end{aligned}$$

Furthermore, we know from Lemma 3.7 (applied to $\hat K$) that $\operatorname{div}\Pi_{\hat K}^k(\hat\tau) = \mathscr{P}_{\hat K}^k(\operatorname{div}\hat\tau)$, where $\mathscr{P}_{\hat K}^k : L^2(\hat K) \to \mathbb{P}_k(\hat K)$ is the orthogonal projector. Then, employing the estimate (3.15) (cf. Lemma 3.12) and the preceding identity, we obtain

$$\begin{aligned}
|\operatorname{div}\tau - \operatorname{div}\Pi_K^k(\tau)|_{m,K} &\leq \hat C\,\|B_K^{-1}\|^m\,|\det B_K|^{1/2}\,|\widehat{\operatorname{div}\tau - \operatorname{div}\Pi_K^k(\tau)}|_{m,\hat K}\\
&= \hat C\,\|B_K^{-1}\|^m\,|\det B_K|^{-1/2}\,|\operatorname{div}\hat\tau - \mathscr{P}_{\hat K}^k(\operatorname{div}\hat\tau)|_{m,\hat K}\,.
\end{aligned} \tag{3.27}$$

Now it is easy to see that $\mathscr{P}_{\hat{K}}^k \in \mathscr{L}(H^{k+1}(\hat{K}), H^m(\hat{K}))$, for instance, writing

$$\mathscr{P}_{\hat{K}}^k(\hat{v}) := \sum_{i=1}^{m_k} \langle \hat{v}, \varphi_{i,k} \rangle_{0,\hat{K}} \varphi_{i,k} \qquad \forall \hat{v} \in L^2(\hat{K}),$$

where $\langle \cdot, \cdot \rangle_{0,\hat{K}}$ is the inner product of $L^2(\hat{K})$ and $\{\varphi_{1,k}, \varphi_{2,k}, \cdots, \varphi_{m_k,k}\}$ is an orthonormal basis of $\mathbb{P}_k(\hat{K})$. In addition, it is clear that $\mathscr{P}_{\hat{K}}^k(\hat{p}) = \hat{p} \quad \forall \hat{p} \in \mathbb{P}_k(\hat{K})$. Hence, employing the Bramble–Hilbert lemma, the identity (3.26), and the estimate (3.14) (cf. Lemma 3.12), we conclude that

$$\begin{aligned} &|\mathrm{div}\, \hat{\tau} - \mathscr{P}_{\hat{K}}^k(\mathrm{div}\, \hat{\tau})|_{m,\hat{K}} \leq C |\mathrm{div}\, \hat{\tau}|_{k+1,\hat{K}} \\ &= C |\det B_K| \, |\widehat{\mathrm{div}\, \tau}|_{k+1,\hat{K}} \leq C |\det B_K|^{1/2} \, \|B_K\|^{k+1} \, |\mathrm{div}\, \tau|_{k+1,K}, \end{aligned} \tag{3.28}$$

which, substituted into (3.27), implies

$$|\mathrm{div}\, \tau - \mathrm{div}\, \Pi_K^k(\tau)|_{m,K} \leq C \|B_K^{-1}\|^m \, \|B_K\|^{k+1} \, |\mathrm{div}\, \tau|_{k+1,K}. \tag{3.29}$$

Finally, using again the geometric bounds given by Lemma 3.14, we obtain (3.22) directly from (3.29).

□

The following result extends Lemma 3.16 to all the intermediate seminorms.

Lemma 3.17. *Let m, k, and l be nonnegative integers such that $0 \leq l \leq k$ and $0 \leq m \leq l+1$. Then there exists $C := C(\hat{K}, \Pi_{\hat{K}}^k, k, m, n) > 0$ such that*

$$|\tau - \Pi_K^k(\tau)|_{m,K} \leq C \frac{h_K^{l+2}}{\rho_K^{m+1}} |\tau|_{l+1,K} \qquad \forall \tau \in [H^{l+1}(K)]^n. \tag{3.30}$$

In addition, for each $\tau \in [H^1(K)]^n$, with $\mathrm{div}\, \tau \in H^{l+1}(K)$, there holds

$$|\mathrm{div}\, \tau - \mathrm{div}\, \Pi_K^k(\tau)|_{m,K} \leq C \frac{h_K^{l+1}}{\rho_K^m} |\mathrm{div}\, \tau|_{l+1,K}. \tag{3.31}$$

Proof. We first observe that the same proof of Lemma 3.15 can be applied here to prove that $\Pi_{\hat{K}}^k \in \mathscr{L}([H^{l+1}(\hat{K})]^n, [H^m(\hat{K})]^n)$. Then, since $\Pi_{\hat{K}}^k(\hat{\mathbf{p}}) = \hat{\mathbf{p}} \quad \forall \hat{\mathbf{p}} \in RT_k(\hat{K})$ and $[\mathbb{P}_l(\hat{K})]^n \subseteq [\mathbb{P}_k(\hat{K})]^n \subseteq RT_k(\hat{K})$, the Bramble–Hilbert lemma implies now, instead of (3.24), that, given $\tau \in [H^{l+1}(K)]^n$, there holds

$$|\hat{\tau} - \Pi_{\hat{K}}^k(\hat{\tau})|_{m,\hat{K}} \leq C |\hat{\tau}|_{l+1,\hat{K}},$$

so that the rest of the derivation of (3.30) is exactly as in the proof of Lemma 3.16.

Next, since $\mathscr{P}_{\hat{K}}^k(\hat{p}) = \hat{p} \quad \forall \hat{p} \in \mathbb{P}_k(\hat{K})$ and $\mathbb{P}_l(\hat{K}) \subseteq \mathbb{P}_k(\hat{K})$, the Bramble–Hilbert lemma implies in this case, instead of (3.28), that, given $\tau \in [H^1(K)]^n$ with $\mathrm{div}\, \tau \in H^{l+1}(K)$, there holds

$$|\mathrm{div}\, \hat{\tau} - \mathscr{P}_{\hat{K}}^k(\mathrm{div}\, \hat{\tau})|_{m,\hat{K}} \leq C |\mathrm{div}\, \hat{\tau}|_{l+1,\hat{K}},$$

and therefore, the rest of the deduction of (3.31) also follows as in the proof of Lemma 3.16.

□

Having estimated the local interpolation error, we are now in a position to estimate the global interpolation error. To this end, we recall that a family of triangularizations $\{\mathscr{T}_h\}_{h>0}$ of $\overline{\Omega}$ is said to be regular if there exists $c > 0$ such that

$$\frac{h_K}{\rho_K} \le c \qquad \forall K \in \mathscr{T}_h, \qquad \forall h > 0. \tag{3.32}$$

We have the following main result.

Theorem 3.6 (Global Interpolation Error). *Let $\{\mathscr{T}_h\}_{h>0}$ be a regular family of triangularizations of $\overline{\Omega}$, and let k be a nonnegative integer. Then there exists $C > 0$, independently of h, such that*

$$\|\tau - \Pi_h^k(\tau)\|_{\mathrm{div},\Omega} \le C h^{l+1} \Big\{ |\tau|_{l+1,\Omega} + |\mathrm{div}\, \tau|_{l+1,\Omega} \Big\} \tag{3.33}$$

for each $\tau \in [H^{l+1}(\Omega)]^n$, with $\mathrm{div}\, \tau \in H^{l+1}(\Omega)$, $0 \le l \le k$.

Proof. Let $0 \le l \le k$ and $\tau \in [H^{l+1}(\Omega)]^n$ such that $\mathrm{div}\, \tau \in H^{l+1}(\Omega)$. Then, applying (3.30) and (3.31) (cf. Lemma 3.17), with $m = 0$, we obtain

$$\|\tau - \Pi_K^k(\tau)\|_{0,K} \le C \frac{h_K^{l+2}}{\rho_K} |\tau|_{l+1,K} \qquad \forall K \in \mathscr{T}_h$$

and

$$\|\mathrm{div}\, \tau - \mathrm{div}\, \Pi_K^k(\tau)\|_{0,K} \le C h_K^{l+1} |\mathrm{div}\, \tau|_{l+1,K} \qquad \forall K \in \mathscr{T}_h,$$

from which, employing the regularity of the family $\{\mathscr{T}_h\}_{h>0}$ [cf. (3.32)], we deduce that

$$\begin{aligned}
\|\tau - \Pi_K^k(\tau)\|_{\mathrm{div},K}^2 &= \|\tau - \Pi_K^k(\tau)\|_{0,K}^2 + \|\mathrm{div}\, \tau - \mathrm{div}\, \Pi_K^k(\tau)\|_{0,K}^2 \\
&\le C^2 c^2 h_K^{2(l+1)} |\tau|_{l+1,K}^2 + C^2 h_K^{2(l+1)} |\mathrm{div}\, \tau|_{l+1,K}^2 \\
&\le \tilde{C}^2 h_K^{2(l+1)} \Big\{ |\tau|_{l+1,K}^2 + |\mathrm{div}\, \tau|_{l+1,K}^2 \Big\},
\end{aligned}$$

where $\tilde{C}^2 := C^2 \max\{c^2, 1\}$.

Then, recalling that $\Pi_h^k(\tau)|_K = \Pi_K^k(\tau|_K)$ and $h_K \le h \, \forall K \in \mathscr{T}_h$, we find that

$$\begin{aligned}
\|\tau - \Pi_h^k(\tau)\|_{\mathrm{div},\Omega}^2 &= \sum_{K \in \mathscr{T}_h} \|\tau - \Pi_K^k(\tau)\|_{\mathrm{div},K}^2 \\
&\le \sum_{K \in \mathscr{T}_h} \tilde{C}^2 h_K^{2(l+1)} \Big\{ |\tau|_{l+1,K}^2 + |\mathrm{div}\, \tau|_{l+1,K}^2 \Big\} \\
&\le \tilde{C}^2 h^{2(l+1)} \Big\{ |\tau|_{l+1,\Omega}^2 + |\mathrm{div}\, \tau|_{l+1,\Omega}^2 \Big\},
\end{aligned}$$

which gives (3.33) and completes the proof.

□

We end this chapter with a couple of additional interpolation error estimates that are very useful in applications.

Lemma 3.18 (Interpolation Error of Normal Components). *There exists $C > 0$, independently of h, such that $\forall\, K \in \mathscr{T}_h$, $\forall$ face/side F of K, and $\forall\, \tau \in [H^1(K)]^n$ there holds*

$$\|\tau \cdot \mathbf{n}_F - \Pi_K^k(\tau)\cdot \mathbf{n}_F\|_{0,F} \le C\,|F|^{1/2}\,|\tau|_{1,K}\,. \tag{3.34}$$

Proof. Let $\hat{F}$ be the face/side of $\hat{K}$ such that $F = T_K(\hat{F})$, and let us define $T_F := T_K|_{\hat{F}}$. Then we have the change of variable formula

$$\int_F f(x)\,ds_x = \frac{|F|}{|\hat{F}|}\int_{\hat{F}} f(T_F(\hat{x}))\,ds_{\hat{x}}\,. \tag{3.35}$$

Now, given $\tau \in [H^1(K)]^n$, we know from Lemma 3.6 and the definition of the operator Π_K^k, respectively, that $\Pi_K^k(\tau)\cdot\mathbf{n}_F \in \mathbb{P}_k(F)$ and

$$\int_F \tau\cdot\mathbf{n}_F\,\psi = \int_F \Pi_K^k(\tau)\cdot\mathbf{n}_F\,\psi \qquad \forall\,\psi\in\mathbb{P}_k(F)\,,$$

which implies that

$$\Pi_K^k(\tau)\cdot\mathbf{n}_F = \mathscr{P}_F^k(\tau\cdot\mathbf{n}_F)\,, \tag{3.36}$$

where $\mathscr{P}_F^k : L^2(F) \to \mathbb{P}_k(F)$ is the orthogonal projector. Then, if $\tilde{}$ denotes composition with T_F, it is easy to see that $\widetilde{\mathscr{P}_F^k(v)} = \mathscr{P}_{\hat{F}}^k(\tilde{v}) \quad \forall\, v\in L^2(F)$, where $\mathscr{P}_{\hat{F}}^k : L^2(\hat{F}) \to \mathbb{P}_k(\hat{F})$ is the corresponding orthogonal projector. Then, utilizing (3.35), we obtain

$$\begin{aligned}
&\|\tau\cdot\mathbf{n}_F - \Pi_K^k(\tau)\cdot\mathbf{n}_F\|_{0,F} = \|\tau\cdot\mathbf{n}_F - \mathscr{P}_F^k(\tau\cdot\mathbf{n}_F)\|_{0,F}\\
&= \frac{|F|^{1/2}}{|\hat{F}|^{1/2}}\,\|\widetilde{\tau\cdot\mathbf{n}_F} - \widetilde{\mathscr{P}_F^k(\tau\cdot\mathbf{n}_F)}\|_{0,\hat{F}}\\
&= \frac{|F|^{1/2}}{|\hat{F}|^{1/2}}\,\|\widetilde{\tau\cdot\mathbf{n}_F} - \mathscr{P}_{\hat{F}}^k(\widetilde{\tau\cdot\mathbf{n}_F})\|_{0,\hat{F}}\\
&\le \frac{|F|^{1/2}}{|\hat{F}|^{1/2}}\,\|\widetilde{\tau\cdot\mathbf{n}_F}\|_{0,\hat{F}} \le \hat{C}\,|F|^{1/2}\,\|\tilde{\tau}\|_{0,\hat{F}} = \hat{C}\,|F|^{1/2}\,\|\tilde{\tau}\,\hat{\varphi}\|_{0,\hat{F}}
\end{aligned}$$

where $\tilde{\tau} := \tau\circ T_K$ in $\hat{K}$ and $\hat{\varphi} \in C^\infty(\hat{K})$ is such that $\hat{\varphi} \equiv 1$ in a neighborhood of $\hat{F}$, and $\hat{\varphi} \equiv 0$ in a neighborhood of the vertex opposite to $\hat{F}$. Thus, starting from the preceding estimate, and applying the trace theorem in $H^1(\hat{K})$, the Friedrichs–Poincaré inequality, the Leibniz rule, the estimate (3.14) (cf. Lemma 3.12) with $m = 1$, and the geometric bounds from Lemma 3.14, we deduce that

$$\|\tau\cdot\mathbf{n}_F - \Pi_K^k(\tau)\cdot\mathbf{n}_F\|_{0,F} \le \hat{C}|F|^{1/2}c\|\tilde{\tau}\hat{\varphi}\|_{1,\hat{K}}$$
$$\le C|F|^{1/2}|\tilde{\tau}\hat{\varphi}|_{1,\hat{K}} \le C|F|^{1/2}|\tilde{\tau}|_{1,\hat{K}}$$
$$\le C|F|^{1/2}\|B_K\|\,|\det B_K|^{-1/2}|\tau|_{1,K}$$
$$\le C|F|^{1/2}\frac{h_K}{\hat{\rho}}\frac{|\hat{K}|^{1/2}}{|K|^{1/2}}|\tau|_{1,K} = \hat{C}|F|^{1/2}|\tau|_{1,K},$$

which completes the proof of (3.34).

□

Furthermore, it is not difficult to prove that the Raviart–Thomas interpolation operator of order 0, that is, Π_h^0, can also be defined in the space $[H^\delta(\Omega)]^n \cap H(\mathrm{div};\Omega)$ $\forall\,\delta \in]0,1[$. Moreover, in this case, we have the following result (cf. S. Meddahi, 2011, private communication).

Lemma 3.19 (Local Interpolation Error with Fractional Order). *Given $\delta \in]0,1[$ and $\tau \in [H^\delta(\Omega)]^n \cap H(\mathrm{div};\Omega)$, there holds*

$$\|\tau - \Pi_K^0(\tau)\|_{0,K} \le C h_K^\delta \left\{ |\tau|_{\delta,K} + \|\mathrm{div}\,\tau\|_{0,K} \right\} \quad \forall K \in \mathscr{T}_h. \tag{3.37}$$

Proof. We begin as in the proof of Lemma 3.16. In fact, employing the estimate (3.19) (cf. Lemma 3.13), the fact that $\Pi_{\hat{K}}^0(\hat{\tau}) = \widehat{\Pi_K^0(\tau)}$ (cf. Lemma 3.11), and the upper bound for $\|B_K\|$ given in Lemma 3.14, we obtain

$$\|\tau - \Pi_K^0(\tau)\|_{0,K} \le \hat{C} h_K |\det B_K|^{-1/2} \|\hat{\tau} - \Pi_{\hat{K}}^0(\hat{\tau})\|_{0,\hat{K}}. \tag{3.38}$$

Next, we know from [44, Eq. (3.39)] that there exists $C > 0$, depending only on $\hat{K}$ and δ, such that

$$\|\Pi_{\hat{K}}^0(\hat{\zeta})\|_{0,\hat{K}} \le C\left\{ \|\hat{\zeta}\|_{\delta,\hat{K}} + \|\mathrm{div}\,\hat{\zeta}\|_{0,\hat{K}} \right\} \quad \forall \hat{\zeta} \in [H^\delta(\hat{K})]^n \cap H(\mathrm{div};\hat{K}). \tag{3.39}$$

Now, it is clear that

$$\|\hat{\tau} - \Pi_{\hat{K}}^0(\hat{\tau})\|_{0,\hat{K}} = \|(\hat{\tau}+\mathbf{p}) - \Pi_{\hat{K}}^0(\hat{\tau}+\mathbf{p})\|_{0,\hat{K}}$$
$$\le \|\hat{\tau}+\mathbf{p}\|_{0,\hat{K}} + \|\Pi_{\hat{K}}^0(\hat{\tau}+\mathbf{p})\|_{0,\hat{K}} \quad \forall \mathbf{p} \in [\mathbb{P}_0(\hat{K})]^n \subseteq RT_0(\hat{K}),$$

which, using (3.39), the estimate $\|\cdot\|_{0,\hat{K}} \le \|\cdot\|_{\delta,\hat{K}}$, and the fact that $\mathrm{div}\,\mathbf{p} = 0$, gives

$$\|\hat{\tau} - \Pi_{\hat{K}}^0(\hat{\tau})\|_{0,\hat{K}} \le C\left\{ \|\hat{\tau}+\mathbf{p}\|_{\delta,\hat{K}} + \|\mathrm{div}\,\hat{\tau}\|_{0,\hat{K}} \right\} \quad \forall \mathbf{p} \in [\mathbb{P}_0(\hat{K})]^n. \tag{3.40}$$

On the other hand, the Deny–Lions lemma (cf. Theorem 3.4) can actually be established for Sobolev spaces of fractional order as well (e.g., [22, Theorem 6.1]), thanks to which there holds, in particular,

$$\inf_{\mathbf{p}\in[\mathbb{P}_0(\hat{K})]^n} \|\mathbf{v}+\mathbf{p}\|_{\delta,\hat{K}} \le C|\mathbf{v}|_{\delta,\hat{K}} \qquad \forall \mathbf{v}\in [H^\delta(\hat{K})]^n, \tag{3.41}$$

where C depends on $\hat{K}$ and δ. We also refer to [43, Lemma 2.3] for this specific estimate (3.41). In this way, taking the infimum with respect to $\mathbf{p}$ in (3.40) and using (3.41), we find that

$$\|\hat{\tau}-\Pi^0_{\hat{K}}(\hat{\tau})\|_{0,\hat{K}} \le C\left\{|\hat{\tau}|_{\delta,\hat{K}} + \|\operatorname{div}\hat{\tau}\|_{0,\hat{K}}\right\},$$

which, inserted back into (3.38), yields

$$\|\tau-\Pi^0_K(\tau)\|_{0,K} \le C h_K |\det B_K|^{-1/2}\left\{|\hat{\tau}|_{\delta,\hat{K}} + \|\operatorname{div}\hat{\tau}\|_{0,\hat{K}}\right\}. \tag{3.42}$$

Next, we notice that the scaling properties of seminorms given by Lemma 3.12 are also valid for fractional Sobolev spaces (e.g., [43, Lemmas 2.8 and 2.9]). Thus, proceeding similarly as in the proof of Lemma 3.13, we obtain that

$$|\hat{\tau}|_{\delta,\hat{K}} \le C\|B_K^{-1}\|\,\|B_K\|^\delta\,|\det B_K|^{1/2}\,|\tau|_{\delta,K}. \tag{3.43}$$

In addition, we recall from (3.13) that $\operatorname{div}\hat{\tau} = |\det B_K|\widehat{\operatorname{div}\tau}$ in $\hat{K}$, which together with (3.14) (cf. Lemma 3.12), implies that

$$\|\operatorname{div}\hat{\tau}\|_{0,\hat{K}} = |\det B_K|\,\|\widehat{\operatorname{div}\tau}\|_{0,\hat{K}} \le C|\det B_K|^{1/2}\,\|\operatorname{div}\tau\|_{0,K}. \tag{3.44}$$

Therefore, gathering (3.43) and (3.44) into (3.42) and employing the upper bounds $\|B_K\| \le \hat{c}_1 h_K$ and $\|B_K^{-1}\| \le \hat{c}_2 h_K^{-1}$ (cf. Lemma 3.14), we deduce that

$$\|\tau-\Pi^0_K(\tau)\|_{0,K} \le C\left\{h_K^\delta\,|\tau|_{\delta,K} + h_K\,\|\operatorname{div}\tau\|_{0,K}\right\},$$

which, noting that $h_K \le h_K^\delta$, provides the required inequality and completes the proof.

□

The definitions and corresponding interpolation and approximation properties of several other finite element subspaces for $H(\operatorname{div};\Omega)$, such as Brezzi–Douglas–Marini (BDM), Brezzi–Douglas–Fortin–Marini (BDFM), and others, can be found in the classic works on the subject (e.g., [13, 16]).

Chapter 4
MIXED FINITE ELEMENT METHODS

In this chapter we utilize the Raviart–Thomas spaces to present and analyze specific mixed finite element methods applied to some of the examples studied in Chap. 2. The corresponding discussion follows mainly the presentations in [12, 39, 50, 52]. We begin with a preliminary section dealing with the approximation properties of the finite element subspaces to be employed.

4.1 Projection Operators

In what follows, Ω is a bounded and connected domain of $\mathbb{R}^n$, $n \in \{2,3\}$, with polyhedral boundary Γ, and $\mathcal{T}_h$ is a triangularization of $\overline{\Omega}$. Then, given a nonnegative integer k (either $k \geq 0$ or $k \geq 1$, as indicated subsequently), we are interested in the following orthogonal projectors (in each case with respect to the inner products of the projected spaces):

$$
\begin{aligned}
\mathscr{P}_{\text{div},h}^k &: H(\text{div};\Omega) \longrightarrow H_h^k, \\
\mathbf{P}_{1,h}^k &: H^1(\Omega) \longrightarrow X_h^k, \\
\mathbf{P}_h^k &: L^2(\Omega) \longrightarrow X_h^k, \\
\mathscr{P}_h^k &: L^2(\Omega) \longrightarrow Y_h^k,
\end{aligned}
$$

where H_h^k is the global Raviart–Thomas space defined at the beginning of Sect. 3.4.1, that is, for each $k \geq 0$

$$
H_h^k := \left\{ \tau_h \in H(\text{div};\Omega) : \quad \tau_h|_K \in RT_k(K) \quad \forall K \in \mathcal{T}_h \right\}, \tag{4.1}
$$

X_h^k is the usual Lagrange finite element space, that is, for each $k \geq 1$

$$
X_h^k := \left\{ v_h \in C(\overline{\Omega}) : \quad v_h|_K \in \mathbb{P}_k(K) \quad \forall K \in \mathcal{T}_h \right\}, \tag{4.2}
$$

G.N. Gatica, *A Simple Introduction to the Mixed Finite Element Method: Theory and Applications*, SpringerBriefs in Mathematics, DOI 10.1007/978-3-319-03695-3_4, © Gabriel N. Gatica 2014

and Y_h^k is the space of piecewise polynomials of degree $k \geq 0$ given by

$$Y_h^k := \left\{ v_h \in L^2(\Omega) : \quad v_h|_K \in \mathbb{P}_k(K) \qquad \forall K \in \mathscr{T}_h \right\}. \tag{4.3}$$

In what follows, we assume that $\mathscr{T}_h$ belongs to a regular family of triangularizations $\{\mathscr{T}_h\}_{h>0}$. In addition, we recall from Chap. 3 [cf. (3.5)] that $\Pi_h^k : H(\mathrm{div};\Omega) \cap Z \to H_h^k$ is the global Raviart–Thomas interpolation operator, where $Z := \left\{ \tau \in [L^2(\Omega)]^n : \quad \tau|_K \in [H^1(K)]^n \quad \forall K \in \mathscr{T}_h \right\}$ (cf. Theorem 3.2). Then we observe that for each $\tau \in H(\mathrm{div};\Omega) \cap Z$ there holds

$$\|\tau - \mathscr{P}_{\mathrm{div},h}^k(\tau)\|_{\mathrm{div},\Omega} := \inf_{\tau_h \in H_h^k} \|\tau - \tau_h\|_{\mathrm{div},\Omega} \leq \|\tau - \Pi_h^k(\tau)\|_{\mathrm{div},\Omega},$$

which, thanks to Theorem 3.6, implies that for each $\tau \in [H^{l+1}(\Omega)]^n$, with $\mathrm{div}\,\tau \in H^{l+1}(\Omega)$, $0 \leq l \leq k$, there also holds

$$\|\tau - \mathscr{P}_{\mathrm{div},h}^k(\tau)\|_{\mathrm{div},\Omega} \leq C h^{l+1} \left\{ |\tau|_{l+1,\Omega} + |\mathrm{div}\,\tau|_{l+1,\Omega} \right\}. \tag{4.4}$$

Then, if $\tilde{\Pi}_h^k : C(\overline{\Omega}) \to X_h^k$ denotes the global Lagrange interpolation operator (e.g., [20, 51]), we obviously find that for each $v \in C(\overline{\Omega})$ there holds

$$\|v - \mathbf{P}_{1,h}^k(v)\|_{1,\Omega} := \inf_{v_h \in X_h^k} \|v - v_h\|_{1,\Omega} \leq \|v - \tilde{\Pi}_h^k(v)\|_{1,\Omega}$$

$$\leq \|v - \tilde{\Pi}_h^k(v)\|_{0,\Omega} + |v - \tilde{\Pi}_h^k(v)|_{1,\Omega}$$

and

$$\|v - \mathbf{P}_h^k(v)\|_{0,\Omega} := \inf_{v_h \in X_h^k} \|v - v_h\|_{0,\Omega} \leq \|v - \tilde{\Pi}_h^k(v)\|_{0,\Omega},$$

which, thanks to the known estimates for $\tilde{\Pi}_h^k$ (e.g., [20, 51]), yields for each $v \in H^{l+1}(\Omega)$, $1 \leq l \leq k$, that

$$\|v - \mathbf{P}_{1,h}^k(v)\|_{1,\Omega} \leq C h^{l+1} |v|_{l+1,\Omega} + C h^l |v|_{l+1,\Omega} \leq C h^l |v|_{l+1,\Omega}, \tag{4.5}$$

and

$$\|v - \mathbf{P}_h^k(v)\|_{0,\Omega} \leq C h^{l+1} |v|_{l+1,\Omega}. \tag{4.6}$$

Now, applying the Bramble–Hilbert lemma (cf. Theorem 3.5) to $S = \Omega$ and $\Pi := \mathbf{P}_{1,h}^k$ (with $m = 1$ and $k = 0$), and noting that $\Pi(p) = p \quad \forall p \in \mathbb{P}_0(\Omega) \subseteq X_h^k$, we deduce that

$$\|v - \mathbf{P}_{1,h}^k(v)\|_{1,\Omega} \leq C |v|_{1,\Omega},$$

which proves that (4.5) can be extended to $l = 0$, and therefore we can write

$$\|v - \mathbf{P}_{1,h}^k(v)\|_{1,\Omega} \leq C h^l |v|_{l+1,\Omega} \qquad \forall v \in H^{l+1}(\Omega), \quad 0 \leq l \leq k. \tag{4.7}$$

In addition, (4.6) can also be extended to $l = 0$ (with $\|\cdot\|_{1,\Omega}$ instead of $|\cdot|_{1,\Omega}$ on the right-hand side), but this extension requires some results on the interpolation of Sobolev spaces (cf. [49, Appendix B]). Nevertheless, later on we give an alternative deduction of this estimate (with $|\cdot|_{1,\Omega}$), which, as a consequence of the result to be presented next, is valid only for a convex domain Ω [see (4.14)].

More precisely, the following lemma, known as the *Aubin–Nitsche trick*, assumes that the domain Ω is convex and utilizes a duality argument to establish an estimate of the projection error $\mathbf{I} - \mathbf{P}_{1,h}^k$ measured in the norm $\|\cdot\|_{0,\Omega}$. Hereafter, $\mathbf{I}$ denotes a generic identity operator.

Lemma 4.1. *Let Ω be a convex domain, and let $k \geq 1$. Then there exists $C > 0$, independently of h, such that for each $v \in H^{l+1}(\Omega)$, $0 \leq l \leq k$, there holds*

$$\|v - \mathbf{P}_{1,h}^k(v)\|_{0,\Omega} \leq C h^{l+1} |v|_{l+1,\Omega}. \tag{4.8}$$

Proof. Let $T : L^2(\Omega) \to H^1(\Omega)$ be the linear and bounded operator assigning to each $r \in L^2(\Omega)$ a unique solution $T(r) \in H^1(\Omega)$, which is guaranteed by the Riesz representation theorem, of the problem

$$\langle T(r), w\rangle_{1,\Omega} = \langle r, w\rangle_{0,\Omega} \qquad \forall w \in H^1(\Omega), \tag{4.9}$$

where $\langle\cdot,\cdot\rangle_{1,\Omega}$ and $\langle\cdot,\cdot\rangle_{0,\Omega}$ denote the inner products of $H^1(\Omega)$ and $L^2(\Omega)$, respectively. Note that (4.9) is the variational formulation of the boundary value problem

$$-\Delta T(r) + T(r) = r \quad \text{in} \quad \Omega, \quad \nabla T(r)\cdot\mathbf{n} = 0 \quad \text{on} \quad \partial\Omega,$$

where $\mathbf{n}$ is the normal vector at $\partial\Omega$. Since Ω is convex, the corresponding elliptic regularity result says that $T(r) \in H^2(\Omega)$ and that there exists $C := C(\Omega) > 0$ such that

$$\|T(r)\|_{2,\Omega} \leq C\|r\|_{0,\Omega} \qquad \forall r \in L^2(\Omega). \tag{4.10}$$

Now, given $v \in H^{l+1}(\Omega)$, $0 \leq l \leq k$, we obtain, utilizing (4.9), that

$$\begin{aligned}
\|v - \mathbf{P}_{1,h}^k(v)\|_{0,\Omega} &= \sup_{\substack{r\in L^2(\Omega)\\ r\neq 0}} \frac{\langle r, v - \mathbf{P}_{1,h}^k(v)\rangle_{0,\Omega}}{\|r\|_{0,\Omega}} \\
&= \sup_{\substack{r\in L^2(\Omega)\\ r\neq 0}} \frac{\langle T(r), v - \mathbf{P}_{1,h}^k(v)\rangle_{1,\Omega}}{\|r\|_{0,\Omega}} \\
&= \sup_{\substack{r\in L^2(\Omega)\\ r\neq 0}} \frac{\langle T(r) - v_h, v - \mathbf{P}_{1,h}^k(v)\rangle_{1,\Omega}}{\|r\|_{0,\Omega}}
\end{aligned}$$

for each $v_h \in X_h^k$, where the orthogonality condition characterizing the orthogonal projector $\mathbf{P}_{1,h}^k$ is used in the last equality. In particular, taking $v_h := \mathbf{P}_{1,h}^k(T(r))$ and applying the Cauchy–Schwarz inequality, we find that

$$\|v - \mathbf{P}_{1,h}^k(v)\|_{0,\Omega} \le \|v - \mathbf{P}_{1,h}^k(v)\|_{1,\Omega} \sup_{\substack{r \in L^2(\Omega) \\ r \neq 0}} \frac{\|T(r) - \mathbf{P}_{1,h}^k(T(r))\|_{1,\Omega}}{\|r\|_{0,\Omega}}. \tag{4.11}$$

Moreover, since $T(r) \in H^2(\Omega)$, a direct application of (4.5) (with $l = 1$) and the estimate (4.10) imply that

$$\|T(r) - \mathbf{P}_{1,h}^k(T(r))\|_{1,\Omega} \le Ch|T(r)|_{2,\Omega} \le Ch\|r\|_{0,\Omega},$$

which, inserted into (4.11), gives

$$\|v - \mathbf{P}_{1,h}^k(v)\|_{0,\Omega} \le Ch\|v - \mathbf{P}_{1,h}^k(v)\|_{1,\Omega} \tag{4.12}$$

and, using (4.7) (with $l = 0$) in (4.12), yields

$$\|v - \mathbf{P}_{1,h}^k(v)\|_{0,\Omega} \le Ch|v|_{1,\Omega} \qquad \forall v \in H^1(\Omega). \tag{4.13}$$

Finally, combining (4.12) and (4.5) we conclude that

$$\|v - \mathbf{P}_{1,h}^k(v)\|_{0,\Omega} \le Ch^{l+1}|v|_{l+1,\Omega} \qquad \forall v \in H^{l+1}(\Omega),\ 1 \le l \le k.$$

This last estimate, together with (4.13), gives (4.8) and completes the proof. □

It is important to remark here, as stated earlier, that starting from the preceding lemma one can extend inequality (4.6) to $l = 0$. In fact, it is clear that

$$\|v - \mathbf{P}_h^k(v)\|_{0,\Omega} \le \|v - \mathbf{P}_{1,h}^k(v)\|_{0,\Omega},$$

from which, using (4.8) with $l = 0$, it follows that

$$\|v - \mathbf{P}_h^k(v)\|_{0,\Omega} \le Ch|v|_{1,\Omega} \qquad \forall v \in H^1(\Omega), \tag{4.14}$$

and therefore (4.6) can be replaced by

$$\|v - \mathbf{P}_h^k(v)\|_{0,\Omega} \le Ch^{l+1}|v|_{l+1,\Omega} \qquad \forall v \in H^{l+1}(\Omega), \quad 0 \le l \le k. \tag{4.15}$$

The next goal is to prove, under suitable hypotheses, that the projection errors $\mathbf{I} - \mathbf{P}_h^k$ and $\mathbf{I} - \mathbf{P}_{1,h}^k$, both measured in $\|\cdot\|_{1,\Omega}$, are equivalent. Indeed, it is easy to see first that

$$\|v - \mathbf{P}_{1,h}^k(v)\|_{1,\Omega} := \inf_{v_h \in X_h^k} \|v - v_h\|_{1,\Omega} \le \|v - \mathbf{P}_h^k(v)\|_{1,\Omega} \qquad \forall v \in H^1(\Omega). \tag{4.16}$$

To complete this equivalence, an inverse inequality satisfied by the elements of X_h^k is required. To this end, we now assume that the family $\{\mathscr{T}_h\}_{h>0}$ is quasi-uniform, which means that, besides being regular [cf. (3.32)], there exists $\tilde{c} > 0$ such that

$$\min_{K \in \mathscr{T}_h} h_K \ge \tilde{c}h \qquad \forall h > 0.$$

Then we have the following result.

Lemma 4.2. *There exists $\tilde{C} > 0$, independent of h, such that*

$$|v_h|_{1,\Omega} \leq \tilde{C} h^{-1} \|v_h\|_{0,\Omega} \quad \forall v_h \in X_h^k.$$

Proof. Given $v_h \in X_h^k$ we can obviously write

$$|v_h|_{1,\Omega}^2 = \sum_{K \in \mathscr{T}_h} |v_h|_{1,K}^2. \tag{4.17}$$

Then, applying estimate (3.15) (cf. Lemma 3.12) and the geometric bounds given by Lemma 3.14, we obtain

$$\begin{aligned} |v_h|_{1,K} &\leq C |\det B_K|^{1/2} \|B_K^{-1}\| \, |\hat{v}_h|_{1,\hat{K}} \\ &\leq C |\det B_K|^{1/2} \frac{\hat{h}}{\rho_K} |\hat{v}_h|_{1,\hat{K}} \\ &= C |\det B_K|^{1/2} \left(\frac{h_K}{\rho_K}\right) \hat{h} h_K^{-1} |\hat{v}_h|_{1,\hat{K}}, \end{aligned}$$

from which, using that $\frac{h_K}{\rho_K} \leq c$ [cf. (3.32)] and $h_K \geq \tilde{c} h \quad \forall K \in \mathscr{T}_h$, it follows that

$$|v_h|_{1,K} \leq C |\det B_K|^{1/2} h^{-1} |\hat{v}_h|_{1,\hat{K}}.$$

But, since the norms are certainly equivalent in the finite-dimensional space $\mathbb{P}_k(\hat{K})$, there holds $|\hat{v}_h|_{1,\hat{K}} \leq C \|\hat{v}_h\|_{0,\hat{K}}$, and hence, using now estimate (3.14) (cf. Lemma 3.12), we deduce that

$$\begin{aligned} |v_h|_{1,K} &\leq C |\det B_K|^{1/2} h^{-1} \|\hat{v}_h\|_{0,\hat{K}} \\ &\leq C |\det B_K|^{1/2} h^{-1} \|B_K\|^{\circ} |\det B_K|^{-1/2} \|v_h\|_{0,K} \\ &= C h^{-1} \|v_h\|_{0,K}. \end{aligned} \tag{4.18}$$

In this way, (4.17) and (4.18) provide the required inequality. □

We are now in a position to prove the missing inequality in (4.16).

Lemma 4.3. *There exists $C > 0$, independent of h, such that*

$$\|v - \mathbf{P}_h^k(v)\|_{1,\Omega} \leq C \|v - \mathbf{P}_{1,h}^k(v)\|_{1,\Omega} \qquad \forall v \in H^1(\Omega).$$

Proof. Let us observe first that, given $v \in H^1(\Omega)$, we have

$$\begin{aligned} \|v - \mathbf{P}_h^k(v)\|_{1,\Omega} &\leq \|v - \mathbf{P}_{1,h}^k(v)\|_{1,\Omega} + \|\mathbf{P}_{1,h}^k(v) - \mathbf{P}_h^k(v)\|_{1,\Omega} \\ &= \|v - \mathbf{P}_{1,h}^k(v)\|_{1,\Omega} + \|\mathbf{P}_h^k(v - \mathbf{P}_{1,h}^k(v))\|_{1,\Omega}, \end{aligned} \tag{4.19}$$

where we have used that $\mathbf{P}_{1,h}^k(v) = \mathbf{P}_h^k(\mathbf{P}_{1,h}^k(v))$. Then, applying the inverse inequality given by Lemma 4.2 and the estimate (4.12) (cf. proof of Lemma 4.1), we obtain that

$$\begin{aligned}
\|\mathbf{P}_h^k(v - \mathbf{P}_{1,h}^k(v))\|_{1,\Omega}^2 &= \|\mathbf{P}_h^k(v - \mathbf{P}_{1,h}^k(v))\|_{0,\Omega}^2 + |\mathbf{P}_h^k(v - \mathbf{P}_{1,h}^k(v))|_{1,\Omega}^2 \\
&\leq (1 + \tilde{C}^2 h^{-2}) \|\mathbf{P}_h^k(v - \mathbf{P}_{1,h}^k(v))\|_{0,\Omega}^2 \\
&\leq (1 + \tilde{C}^2 h^{-2}) \|v - \mathbf{P}_{1,h}^k(v)\|_{0,\Omega}^2 \\
&\leq (1 + \tilde{C}^2 h^{-2}) C h^2 \|v - \mathbf{P}_{1,h}^k(v)\|_{1,\Omega}^2 \\
&= C(\tilde{C}^2 + h^2) \|v - \mathbf{P}_{1,h}^k(v)\|_{1,\Omega}^2 \\
&\leq \overline{C} \|v - \mathbf{P}_{1,h}^k(v)\|_{1,\Omega}^2 .
\end{aligned}$$

Hence, the preceding estimate and (4.19) complete the proof. □

According to the foregoing estimates, the approximation properties of the projectors $\mathbf{P}_h^k$ and $\mathbf{P}_{1,h}^k$ are summarized in the following inequality:

$$\left\{ \begin{array}{c} \|v - \mathbf{P}_h^k(v)\|_{0,\Omega} + \|v - \mathbf{P}_{1,h}^k(v)\|_{0,\Omega} \\ + h \|v - \mathbf{P}_h^k(v)\|_{1,\Omega} + h \|v - \mathbf{P}_{1,h}^k(v)\|_{1,\Omega} \end{array} \right\} \leq C h^{l+1} |v|_{l+1,\Omega} \tag{4.20}$$

for each $v \in H^{l+1}(\Omega), \quad 0 \leq l \leq k$.

Note here, in virtue of the preceding analysis, that the estimates that arise from the first term (with $l = 0$), and from the second and third terms on the left-hand side of (4.20), require the convexity of Ω.

Finally, let us consider the projector $\mathcal{P}_h^k : L^2(\Omega) \to Y_h^k$ para $k \geq 0$. In this respect, it is easy to see first that

$$\mathcal{P}_h^k(v)|_K = \mathcal{P}_K^k(v|_K) \qquad \forall v \in L^2(\Omega), \quad \forall K \in \mathcal{T}_h,$$

where $\mathcal{P}_K^k : L^2(K) \to \mathbb{P}_k(K)$ is the local orthogonal projector. Now, from the first term of (4.20) applied to $\Omega = K \in \mathcal{T}_h$, which is obviously convex, we find that

$$\|v - \mathcal{P}_K^k(v)\|_{0,K} \leq C h_K^{l+1} |v|_{l+1,K} \qquad \forall v \in H^{l+1}(K).$$

Then, for each $v \in H^{l+1}(\Omega)$, $0 \leq l \leq k$, there holds

$$\begin{aligned}
\|v - \mathcal{P}_h^k(v)\|_{0,\Omega}^2 &= \sum_{K \in \mathcal{T}_h} \|v - \mathcal{P}_K^k(v)\|_{0,K}^2 \\
&\leq \sum_{K \in \mathcal{T}_h} C^2 h_K^{2(l+1)} |v|_{l+1,K}^2 \leq \overline{C} h^{2(l+1)} |v|_{l+1,\Omega}^2 ,
\end{aligned}$$

that is,

$$\|v - \mathcal{P}_h^k(v)\|_{0,\Omega} \leq C h^{l+1} |v|_{l+1,\Omega} . \tag{4.21}$$

4.2 Poisson Problem

In this section we analyze a Galerkin scheme for the Poisson problem studied in Sect. 2.4.1. For this purpose, we first recall that, given a bounded domain $\Omega \subseteq \mathbb{R}^n, n \in \{2,3\}$, with polyhedral boundary Γ, and given $f \in L^2(\Omega)$ and $g \in H^{1/2}(\Gamma)$, the corresponding mixed variational formulation reduces to [cf. (2.19)] the following problem: find $(\sigma, u) \in H \times Q$ such that

$$\begin{aligned} a(\sigma,\tau) + b(\tau,u) &= F(\tau) \qquad \forall\, \tau \in H\,, \\ b(\sigma,v) \qquad\quad &= G(v) \qquad \forall\, v \in Q\,, \end{aligned}$$

where $H := H(\mathrm{div};\Omega)$, $Q := L^2(\Omega)$, a and b are the bilinear forms defined by

$$a(\sigma,\tau) := \int_\Omega \sigma\cdot\tau \qquad \forall\,(\sigma,\tau) \in H\times H,$$

$$b(\tau,v) := \int_\Omega v\,\mathrm{div}\,\tau \qquad \forall\,(\tau,v) \in H\times Q,$$

and the functionals $F \in H'$ and $G \in Q'$ are given by

$$F(\tau) := \langle \gamma_{\mathbf{n}}(\tau), g\rangle \quad \forall\,\tau\in H, \qquad G(v) := -\int_\Omega f v \quad \forall\, v\in Q.$$

Then, if $\{\mathcal{T}_h\}_{h>0}$ is a regular family of triangularizations of $\overline{\Omega}$ [cf. (3.32)] and k is an integer ≥ 0, we introduce the following finite element spaces [cf. (4.1) and (4.3)]:

$$H_h := H_h^k := \left\{\tau_h \in H(\mathrm{div};\Omega): \quad \tau_h|_K \in RT_k(K) \quad \forall\, K\in\mathcal{T}_h\right\}, \tag{4.22}$$

$$Q_h := Y_h^k := \left\{v_h \in L^2(\Omega): \quad v_h|_K \in \mathbb{P}_k(K) \quad \forall\, K\in\mathcal{T}_h\right\}, \tag{4.23}$$

so that the associated Galerkin scheme is as follows: find $(\sigma_h, u_h) \in H_h \times Q_h$ such that

$$\begin{aligned} a(\sigma_h,\tau_h) + b(\tau_h,u_h) &= F(\tau_h) \qquad \forall\,\tau_h \in H_h\,, \\ b(\sigma_h,v_h) \qquad\quad &= G(v_h) \qquad \forall\, v_h \in Q_h\,. \end{aligned} \tag{4.24}$$

The next goal is to apply the theory developed in Sect. 2.5 to conclude the unique solvability and stability of (4.24). To this end, we observe that V_h, the discrete kernel of b [equivalently, $N(\mathbf{B}_h)$, where $\mathbf{B}_h : H_h \to Q_h$ is the discrete operator induced by b], is given by

$$\begin{aligned} V_h &:= \left\{\tau_h\in H_h: \quad b(\tau_h,v_h) := \int_\Omega v_h\,\mathrm{div}\,\tau_h = 0 \quad \forall\, v_h\in Q_h\right\} \\ &= \left\{\tau_h \in H_h: \quad \mathcal{P}_h^k(\mathrm{div}\,\tau_h) = 0\right\}, \end{aligned}$$

from which, noting precisely that $\operatorname{div}\tau_h \in Y_h^k \ \forall \tau_h \in H_h$, we deduce that

$$V_h := \left\{ \tau_h \in H_h : \quad \operatorname{div}\tau_h = 0 \quad \text{in} \quad \Omega \right\}. \tag{4.25}$$

It follows that for each $\tau_h \in V_h$ there holds

$$a(\tau_h, \tau_h) = \|\tau_h\|_{0,\Omega}^2 = \|\tau_h\|_{\operatorname{div},\Omega}^2,$$

which proves the V_h-ellipticity of a with constant $\tilde{\alpha} = 1$, and hence, according to (2.76), hypothesis (i) of Theorem 2.4 (discrete Babuška–Brezzi theorem) is satisfied.

On the other hand, to establish the discrete inf-sup condition for b [cf. hypothesis (ii) of Theorem 2.4], we need the following previous result.

Lemma 4.4. *Let $\Pi_h^k : H(\operatorname{div};\Omega) \cap Z \longrightarrow H_h^k$ be the global Raviart–Thomas interpolation operator. Then there exists $C > 0$, independent of h, such that*

$$\|\Pi_h^k(\tau)\|_{\operatorname{div},\Omega} \le C\|\tau\|_{1,\Omega} \qquad \forall \tau \in [H^1(\Omega)]^n. \tag{4.26}$$

Proof. Let $\tau \in [H^1(\Omega)]^n$. Then, applying the upper bound for the local interpolation error given by Lemma 3.17 (with $m = 0$ and $l = 0$), and using that $\dfrac{h_K}{\rho_K} \le c$, we obtain

$$\|\tau - \Pi_h^k(\tau)\|_{0,K} \le C\frac{h_K^2}{\rho_K}|\tau|_{1,K} \le \tilde{C}h_K|\tau|_{1,K} \qquad \forall K \in \mathscr{T}_h,$$

and then

$$\begin{aligned}\|\Pi_h^k(\tau)\|_{0,\Omega} &\le \|\tau - \Pi_h^k(\tau)\|_{0,\Omega} + \|\tau\|_{0,\Omega} \\ &\le \tilde{C}h|\tau|_{1,\Omega} + \|\tau\|_{0,\Omega} \le \overline{C}\|\tau\|_{1,\Omega}.\end{aligned} \tag{4.27}$$

Next, since $\operatorname{div}(\Pi_h^k(\tau)) = \mathscr{P}_h^k(\operatorname{div}\tau)$ (cf. Lemma 3.7), there holds

$$\|\operatorname{div}\Pi_h^k(\tau)\|_{0,\Omega} = \|\mathscr{P}_h^k(\operatorname{div}\tau)\|_{0,\Omega} \le \|\operatorname{div}\tau\|_{0,\Omega},$$

which, together with (4.27), implies estimate (4.26). □

We proceed next to prove the existence of a Fortin operator so that we can then apply the corresponding result (cf. Lemma 2.6) and thereby verify the discrete inf-sup condition for b. Specifically, we need to define a family of uniformly bounded operators $\{\Pi_h\}_{h>0} \subseteq \mathscr{L}(H, H_h)$ such that

$$b(\tau - \Pi_h(\tau), v_h) = 0 \qquad \forall \tau \in H, \quad \forall v_h \in Q_h, \quad \forall h > 0.$$

In fact, given $\tau \in H := H(\operatorname{div};\Omega)$, we set

$$f_\tau := \begin{cases} \operatorname{div}\tau & \text{in} \quad \Omega, \\ 0 & \text{in} \quad B\backslash\Omega, \end{cases}$$

where B is an open ball containing $\overline{\Omega}$. Since $f_\tau \in L^2(B)$ and B is obviously convex, the boundary value problem

$$-\Delta z = f_\tau \quad \text{in} \quad B, \quad z = 0 \quad \text{on} \quad \partial B,$$

has a unique solution $z \in H_0^1(B) \cap H^2(B)$ that satisfies

$$\|z\|_{2,B} \leq C\,\|f_\tau\|_{0,B} = C\,\|\operatorname{div}\tau\|_{0,\Omega}.$$

Then we let $\tilde{\tau} := -\nabla z|_\Omega$ and notice that $\tilde{\tau} \in [H^1(\Omega)]^n$, $\operatorname{div}\tilde{\tau} = \operatorname{div}\tau$ in Ω, and

$$\|\tilde{\tau}\|_{1,\Omega} \leq \|z\|_{2,\Omega} \leq \|z\|_{2,B} \leq C\,\|\operatorname{div}\tau\|_{0,\Omega}. \tag{4.28}$$

The foregoing analysis suggests defining the Fortin operator as

$$\Pi_h(\tau) := \Pi_h^k(\tilde{\tau}) \qquad \forall\,\tau \in H(\operatorname{div};\Omega). \tag{4.29}$$

It is important to remark here that the necessity of previously regularizing τ by means of the auxiliary function $\tilde{\tau}$ is explained by the fact that the global Raviart–Thomas interpolation operator is defined not in $H(\operatorname{div};\Omega)$ but in $H(\operatorname{div};\Omega) \cap Z$, which contains the space $[H^1(\Omega)]^n$.

It follows, applying Lemma 4.4 and the estimate (4.28), that

$$\|\Pi_h(\tau)\|_{\operatorname{div},\Omega} = \|\Pi_h^k(\tilde{\tau})\|_{\operatorname{div},\Omega} \leq C\,\|\tilde{\tau}\|_{1,\Omega} \leq C_1\,\|\operatorname{div}\tau\|_{0,\Omega},$$

and then

$$\|\Pi_h(\tau)\|_{\operatorname{div},\Omega} \leq C_1\,\|\tau\|_{\operatorname{div},\Omega} \qquad \forall\,\tau \in H := H(\operatorname{div};\Omega), \tag{4.30}$$

which confirms the uniform boundedness of $\{\Pi_h\}_{h>0}$. Next, using the estimate (3.8) given by Lemma 3.7 and the fact that $\mathscr{P}_h^k$ is the orthogonal projector of $L^2(\Omega)$ into $Q_h := Y_h^k$, we deduce that for each $\tau \in H$ and for each $v_h \in Q_h$ there holds

$$\begin{aligned}
b(\tau - \Pi_h(\tau), v_h) &= \int_\Omega v_h\,(\operatorname{div}\tau - \operatorname{div}\Pi_h(\tau)) \\
&= \int_\Omega v_h\,(\operatorname{div}\tau - \operatorname{div}\Pi_h^k(\tilde{\tau})) \\
&= \int_\Omega v_h\,(\operatorname{div}\tau - \mathscr{P}_h^k(\operatorname{div}\tilde{\tau})) \\
&= \int_\Omega v_h\,(\operatorname{div}\tau - \mathscr{P}_h^k(\operatorname{div}\tau)) = 0,
\end{aligned}$$

which yields the second property required by $\{\Pi_h\}_{h>0}$, and consequently the Fortin lemma (cf. Lemma 2.6) guarantees that b satisfies the discrete inf-sup condition on $H_h \times Q_h$ with a constant $\tilde{\beta} > 0$, independently of h. It is important to observe here that, proceeding similarly to the construction of the present Fortin operator, one can prove that in this case there also holds $\operatorname{div} H_h = Q_h$. Indeed, given $v_h \in Q_h$, it suffices to replace $\operatorname{div} \tau$ by v_h in the preceding definition of f_τ, which leads to $\operatorname{div} \Pi_h^k(\tilde{\tau}) = \mathscr{P}_h^k(\operatorname{div} \tilde{\tau}) = \mathscr{P}_h^k(v_h) = v_h$ in Ω.

Consequently, a straightforward application of Theorems 2.4 and 2.6 implies that there exist a unique $(\sigma_h, u_h) \in H_h \times Q_h$ solution of (4.22) and constants $C_1, C_2 > 0$, independent of h, such that

$$\|(\sigma_h, u_h)\|_{H\times Q} \leq C_1 \left\{ \|f\|_{0,\Omega} + \|g\|_{1/2,\Gamma} \right\}$$

and

$$\|\sigma - \sigma_h\|_H + \|u - u_h\|_Q \leq C_2 \left\{ \operatorname{dist}(\sigma, H_h) + \operatorname{dist}(u, Q_h) \right\}, \tag{4.31}$$

where C_1 depends on $\|\mathbf{A}\| \leq 1$ (norm of the operator induced by a), $\tilde{\alpha} = 1$, and $\tilde{\beta}$, whereas C_2 depends on $\|\mathbf{A}\|$, $\|\mathbf{B}\| \leq 1$ (norm of the operator induced by b), $\tilde{\alpha}$, and $\tilde{\beta}$.

Now, according to the upper bounds for the projection errors given by (4.4) and (4.21), we have, respectively,

$$\operatorname{dist}(\sigma, H_h) := \|\sigma - \mathscr{P}_{\operatorname{div},h}^k(\sigma)\|_{\operatorname{div},\Omega} \leq C h^{l+1} \left\{ |\sigma|_{l+1,\Omega} + |\operatorname{div} \sigma|_{l+1,\Omega} \right\} \tag{4.32}$$

if $\sigma \in [H^{l+1}(\Omega)]^n$, with $\operatorname{div} \sigma \in H^{l+1}(\Omega)$, $0 \leq l \leq k$, and

$$\operatorname{dist}(u, Q_h) := \|u - \mathscr{P}_h^k(u)\|_{0,\Omega} \leq C h^{l+1} \, |u|_{l+1,\Omega} \tag{4.33}$$

if $u \in H^{l+1}(\Omega)$, $0 \leq l \leq k$. Therefore, under these regularity assumptions on the exact solution $(\sigma, u) \in H \times Q$, we deduce that the rate of convergence of the Galerkin method (4.24) is given by the estimate that follows from (4.31)–(4.33), that is,

$$\|\sigma - \sigma_h\|_{\operatorname{div},\Omega} + \|u - u_h\|_{0,\Omega} \leq C h^{l+1} \left\{ |\sigma|_{l+1,\Omega} + |\operatorname{div} \sigma|_{l+1,\Omega} + |u|_{l+1,\Omega} \right\}.$$

On the other hand, if (σ, u) is not sufficiently regular and estimates (4.32) and (4.33) do not necessarily hold, the convergence of the Galerkin scheme (4.24), but without any rate of convergence, can still be proved by employing suitable density arguments. More precisely, we have the following result.

Lemma 4.5. *Let $(\sigma, u) \in H \times Q$ and $(\sigma_h, u_h) \in H_h \times Q_h$ be the solutions of the continuous and discrete formulations, respectively. Then*

$$\lim_{h\to 0} \left\{ \|\sigma - \sigma_h\|_{\operatorname{div},\Omega} + \|u - u_h\|_{0,\Omega} \right\} = 0. \tag{4.34}$$

Proof. We use that $[C^\infty(\overline{\Omega})]^n$ and $C_0^\infty(\Omega)$ are dense in $H(\operatorname{div};\Omega)$ and $L^2(\Omega)$, respectively, whence there exist sequences $\{\sigma_j\}_{j\in\mathbb{N}} \subseteq [C^\infty(\overline{\Omega})]^n$ and $\{u_j\}_{j\in\mathbb{N}} \subseteq C_0^\infty(\Omega)$ such that $\|\sigma - \sigma_j\|_{\operatorname{div},\Omega} \overset{j\to\infty}{\longrightarrow} 0$ and $\|u - u_j\|_{0,\Omega} \overset{j\to\infty}{\longrightarrow} 0$. Then, since it is clear that $\sigma_j \in [H^1(\Omega)]^n$, $\operatorname{div} \sigma_j \in H^1(\Omega)$, and $u_j \in H^1(\Omega)$, it follows from (4.4) and (4.21) that for each $j \in \mathbb{N}$

$$\|\sigma_j - \mathscr{P}^k_{\mathrm{div},h}(\sigma_j)\|_{\mathrm{div},\Omega} \le Ch\{|\sigma_j|_{1,\Omega} + |\mathrm{div}\,\sigma_j|_{1,\Omega}\} \tag{4.35}$$

and

$$\|u_j - \mathscr{P}^k_h(u_j)\|_{0,\Omega} \le Ch|u_j|_{1,\Omega}. \tag{4.36}$$

Now, given $\varepsilon > 0$, there exists $N \in \mathbb{N}$ such that

$$\|\sigma - \sigma_N\|_{\mathrm{div},\Omega} < \varepsilon/4 \quad \text{and} \quad \|u - u_N\|_{0,\Omega} < \varepsilon/4.$$

Next, for $j = N$ we deduce from (4.35) and (4.36) that there exists $h_0 > 0$ such that

$$\|\sigma_N - \mathscr{P}^k_{\mathrm{div},h}(\sigma_N)\|_{\mathrm{div},\Omega} < \varepsilon/4 \quad \text{and} \quad \|u_N - \mathscr{P}^k_h(u_N)\|_{0,\Omega} < \varepsilon/4 \qquad \forall h \le h_0.$$

Therefore, from the Cea estimate (4.31) we conclude that for each $h \le h_0$

$$\begin{aligned}
\|\sigma - \sigma_h\|_{\mathrm{div},\Omega} + \|u - u_h\|_{0,\Omega} &\le C_2\Big\{\mathrm{dist}(\sigma, H_h) + \mathrm{dist}(u, Q_h)\Big\}\\
&\le C_2\Big\{\|\sigma - \mathscr{P}^k_{\mathrm{div},h}(\sigma_N)\|_{\mathrm{div},\Omega} + \|u - \mathscr{P}^k_h(u_N)\|_{0,\Omega}\Big\}\\
&\le C_2\Big\{\|\sigma - \sigma_N\|_{\mathrm{div},\Omega} + \|\sigma_N - \mathscr{P}^k_{\mathrm{div},h}(\sigma_N)\|_{\mathrm{div},\Omega}\\
&\quad + \quad \|u - u_N\|_{0,\Omega} + \|u_N - \mathscr{P}^k_h(u_N)\|_{0,\Omega}\Big\}\\
&\le C_2\,\varepsilon,
\end{aligned}$$

which proves the convergence (4.34). □

4.3 Primal-Mixed Formulation of Poisson Problem

In this section we analyze a Galerkin scheme for the primal-mixed formulation of the two-dimensional version of the Poisson problem studied in Sect. 2.4.4. To this end, we recall that, given a bounded domain $\Omega \subseteq \mathbb{R}^2$ with polygonal boundary Γ, and given data $f \in L^2(\Omega)$ and $g \in H^{1/2}(\Gamma)$, the primal-mixed formulation reduces to [cf. (2.70)] the following: find $(u, \xi) \in H \times Q$ such that

$$\begin{aligned}
a(u,v) + b(v,\xi) &= F(v) \qquad \forall v \in H,\\
b(u,\lambda) \qquad\quad &= G(\lambda) \qquad \forall \lambda \in Q,
\end{aligned}$$

where $H := H^1(\Omega)$, $Q := H^{-1/2}(\Gamma)$, a and b are the bilinear forms defined by

$$\begin{aligned}
a(u,v) &:= \int_\Omega \nabla u \cdot \nabla v \quad \forall (u,v) \in H \times H,\\
b(v,\lambda) &:= \langle \lambda, v\rangle \quad \forall (v,\lambda) \in H \times Q,
\end{aligned}$$

and the functionals $F \in H'$ and $G \in Q$ are given by

$$F(v) := \int_\Omega f v \quad \forall v \in H, \qquad G(\lambda) = \langle \lambda, g \rangle \quad \forall \lambda \in Q.$$

Then, given a regular family of triangularizations $\{\mathscr{T}_h\}_{h>0}$ of $\overline{\Omega}$, we introduce the subspaces of H [cf. (4.21)] and Q:

$$H_h := X_h^1 := \left\{ v \in C(\overline{\Omega}) : \quad v|_K \in \mathbb{P}_1(K) \qquad \forall K \in \mathscr{T}_h \right\},$$

$$Q_{\tilde{h}} := \left\{ \lambda \in L^2(\Gamma) : \quad \lambda|_{\tilde{\Gamma}_j} \in \mathbb{P}_0(\tilde{\Gamma}_j) \qquad \forall j \in \{1, \cdots, m\} \right\},$$

where $\{\tilde{\Gamma}_1, \tilde{\Gamma}_2, \cdots, \tilde{\Gamma}_m\}$ is a partition of Γ (independent of the one inherited from $\mathscr{T}_h$) and $\tilde{h} := \max \left\{ |\tilde{\Gamma}_j| : j \in \{1, \cdots, m\} \right\}$. Hence, the associated Galerkin scheme is as follows: find $(u_h, \xi_{\tilde{h}}) \in H_h \times Q_{\tilde{h}}$ such that

$$\begin{aligned} a(u_h, v_h) + b(v_h, \xi_{\tilde{h}}) &= F(v_h) \qquad \forall v_h \in H_h, \\ b(u_h, \lambda_{\tilde{h}}) &= G(\lambda_{\tilde{h}}) \qquad \forall \lambda_{\tilde{h}} \in Q_{\tilde{h}}. \end{aligned} \tag{4.37}$$

We now let V_h be the discrete kernel of b, that is,

$$\begin{aligned} V_h &:= \left\{ v_h \in H_h : \quad b(v_h, \lambda_{\tilde{h}}) = 0 \qquad \forall \lambda_{\tilde{h}} \in Q_{\tilde{h}} \right\} \\ &= \left\{ v_h \in H_h : \quad \langle \lambda_{\tilde{h}}, v_h \rangle \quad = 0 \qquad \forall \lambda_{\tilde{h}} \in Q_{\tilde{h}} \right\}. \end{aligned}$$

Note that, in particular, $\lambda_{\tilde{h}} \equiv 1$ belongs to $Q_{\tilde{h}}$, and therefore

$$V_h \subseteq \hat{V} := \left\{ v \in H : \quad \langle 1, v \rangle = 0 \right\},$$

that is,

$$V_h \subseteq \hat{V} := \left\{ v \in H : \int_\Gamma v = 0 \right\}.$$

Then, utilizing the generalized Poincaré inequality (cf. [46, Theorem 5.11.2]), one can prove that $\|\cdot\|_{1,\Omega}$ and $|\cdot|_{1,\Omega}$ are equivalent in $\hat{V}$ and, hence, in V_h. It follows that

$$a(v_h, v_h) = |v_h|_{1,\Omega} \geq c \, \|v_h\|_{1,\Omega} \qquad \forall v_h \in V_h,$$

which proves that a is V_h-elliptic.

We prove next that b satisfies the discrete inf-sup condition, that is, there exists $\beta > 0$, independent of h, such that

$$\sup_{\substack{v_h \in H_h \\ v_h \neq 0}} \frac{b(v_h, \lambda_{\tilde{h}})}{\|v_h\|_{1,\Omega}} \geq \beta \, \|\lambda_{\tilde{h}}\|_{-1/2,\Gamma} \qquad \forall \lambda_{\tilde{h}} \in Q_{\tilde{h}},$$

which is

$$\sup_{\substack{v_h \in H_h \\ v_h \neq 0}} \frac{\langle \lambda_{\tilde{h}}, v_h \rangle}{\|v_h\|_{1,\Omega}} \geq \beta \, \|\lambda_{\tilde{h}}\|_{-1/2,\Gamma} \qquad \forall \lambda_{\tilde{h}} \in Q_{\tilde{h}}.$$

For this purpose we need an inverse inequality for $Q_{\tilde{h}}$, which is proved by the following lemma for a generic space Q_h.

Lemma 4.6. *Let $\{\Gamma_1, \Gamma_2, \cdots, \Gamma_m\}$ be a partition of Γ, denote $h_j := |\Gamma_j| \quad \forall j \in \{1, \cdots, m\}$, assume that there exists $c > 0$ such that*

$$h_j \geq c\,h := c \max_{i \in \{1,\cdots,m\}} h_i \qquad \forall j \in \{1, \cdots, m\},$$

and define

$$Q_h := \Big\{ \lambda_h \in L^2(\Gamma) : \quad \lambda_h|_{\Gamma_j} \in \mathbb{P}_0(\Gamma_j) \quad \forall j \in \{1, \cdots, m\} \Big\}.$$

Then there exists $C > 0$ such that

$$\|\lambda_h\|_{r,\Gamma} \leq C h^{-1/2-r} \, \|\lambda_h\|_{-1/2,\Gamma} \qquad \forall \lambda_h \in Q_h, \quad \forall r \in [-1/2, 0].$$

Proof. Since clearly

$$\|\lambda_h\|_{-1/2,\Gamma} \leq h^0 \, \|\lambda_h\|_{-1/2,\Gamma} \quad \forall \lambda_h \in Q_h \,,$$

it suffices to prove that

$$\|\lambda_h\|_{0,\Gamma} \leq C h^{-1/2} \, \|\lambda_h\|_{-1/2,\Gamma} \quad \forall \lambda_h \in Q_h \,, \tag{4.38}$$

and then conclude by interpolation estimates (cf. [49, Appendix B]). In fact, given $\lambda_h \in Q_h$, we let $\lambda_j := \lambda_h|_{\Gamma_j} \in \mathbb{P}_0(\Gamma_j)$ and observe that

$$\|\lambda_h\|^2_{0,\Gamma} = \sum_{j=1}^{m} \|\lambda_h\|^2_{0,\Gamma_j} = \sum_{j=1}^{m} h_j \lambda_j^2 = \sum_{j=1}^{m} h_j \, \|\lambda_j\|^2_{0,\hat{\Gamma}},$$

where $\hat{\Gamma}$ is a reference segment of measure $|\hat{\Gamma}| = 1$. For instance, we can consider $\hat{\Gamma} := \{(x,0) : \quad x \in]0,1[\}$. Then, using the equivalence of norms in finite dimension, we have that

$$\|\lambda_h\|^2_{0,\Gamma} \leq \hat{c} \sum_{j=1}^{m} h_j \, \|\lambda_j\|^2_{-1/2,00,\hat{\Gamma}}, \tag{4.39}$$

where $\|\cdot\|_{-1/2,00,\hat{\Gamma}}$ is the norm of $H_{00}^{-1/2}(\hat{\Gamma})$, which is the dual of $H_{00}^{1/2}(\hat{\Gamma})$. On the other hand, applying the inequality [cf. (3.15), Lemma 3.12]

$$|v|_{m,K} \leq c \, \|B_K^{-1}\|^m \, |\det B_K|^{1/2} \, |\hat{v}|_{m,\hat{K}}$$

to $K = \Gamma_j$ and $\hat{K} = \hat{\Gamma}$, we obtain

$$\|v\|_{0,\Gamma_j} \le c \left\{ \frac{|\Gamma_j|}{|\hat{\Gamma}|} \right\}^{1/2} \|\hat{v}\|_{0,\hat{\Gamma}} = \hat{c}\, h_j^{1/2} \|\hat{v}\|_{0,\hat{\Gamma}} \qquad \forall v \in L^2(\Gamma_j)$$

and

$$|v|_{1,\Gamma_j} \le c \left\{ \frac{\hat{h}}{\rho_j} \right\}^{1} \left\{ \frac{|\Gamma_j|}{|\hat{\Gamma}|} \right\}^{1/2} |\hat{v}|_{1,\hat{\Gamma}} \le \hat{c}\, h_j^{-1/2} |\hat{v}|_{1,\hat{\Gamma}} \qquad \forall v \in H^1(\Gamma_j).$$

Then, according to the interpolation estimates for Sobolev spaces (cf. [49, Appendix B]) and using that $H_{00}^{1/2}(\Gamma_j) = (H_0^1(\Gamma_j), L^2(\Gamma_j))_{1/2}$, we find that

$$\|v\|_{1/2,00,\Gamma_j} \le \hat{c}\, \|\hat{v}\|_{1/2,00,\hat{\Gamma}} \qquad \forall v \in H_{00}^{1/2}(\Gamma_j),$$

where $\|\cdot\|_{1/2,00,S}$ denotes the norm of $H_{00}^{1/2}(S)$ for $S \in \{\Gamma_j, \hat{\Gamma}\}$.

Analogously, applying now [cf. (3.14), Lemma 3.12]

$$|\hat{v}|_{m,\hat{K}} \le c\, \|B_K\|^m \, |\det B_K|^{-1/2} \, |v|_{m,K}$$

to $K = \Gamma_j$ and $\hat{K} = \hat{\Gamma}$, using interpolation estimates again, and noting that $H_{00}^{1/2}(\hat{\Gamma})$ is given by $(H_0^1(\hat{\Gamma}), L^2(\hat{\Gamma}))_{1/2}$, we deduce that

$$\|\hat{v}\|_{1/2,00,\hat{\Gamma}} \le \hat{c}\, \|v\|_{1/2,00,\Gamma_j} \qquad \forall \hat{v} \in H_{00}^{1/2}(\hat{\Gamma}),$$

and therefore

$$\|v\|_{1/2,00,\Gamma_j} \cong \|\hat{v}\|_{1/2,00,\hat{\Gamma}}.$$

Consequently, given $\lambda \in H_{00}^{-1/2}(\Gamma_j)$, we obtain by a duality argument that

$$\begin{aligned}
\|\hat{\lambda}\|_{-1/2,00,\hat{\Gamma}} &= \sup_{\substack{\hat{v} \in H_{00}^{1/2}(\hat{\Gamma}) \\ \hat{v} \ne 0}} \frac{\langle \hat{\lambda}, \hat{v} \rangle_{\hat{\Gamma}}}{\|\hat{v}\|_{1/2,00,\hat{\Gamma}}} \\
&= \sup_{\substack{\hat{v} \in H_{00}^{1/2}(\hat{\Gamma}) \\ \hat{v} \ne 0}} \frac{h_j^{-1} \langle \lambda, v \rangle_{\Gamma_j}}{\|\hat{v}\|_{1/2,00,\hat{\Gamma}}} \\
&\le C \sup_{\substack{\hat{v} \in H_{00}^{1/2}(\hat{\Gamma}) \\ \hat{v} \ne 0}} \frac{h_j^{-1} \|\lambda\|_{-1/2,00,\Gamma_j} \|v\|_{1/2,00,\Gamma_j}}{\|v\|_{1/2,00,\Gamma_j}} \\
&= C h_j^{-1} \|\lambda\|_{-1/2,00,\Gamma_j}.
\end{aligned}$$

In this way, employing the preceding estimate in (4.39), we conclude that

$$\|\lambda_h\|_{0,\Gamma}^2 \le \hat{c} \sum_{j=1}^{m} h_j h_j^{-2} \|\lambda_j\|_{-1/2,00,\Gamma_j}^2$$

$$\le \hat{c} h^{-1} \sum_{j=1}^{m} \|\lambda_j\|_{-1/2,00,\Gamma_j}^2 \le \hat{c} h^{-1} \|\lambda_h\|_{-1/2,\Gamma}^2,$$

which gives (4.38) and completes the proof. □

We are now in a position to prove the discrete inf-sup condition for b.

Lemma 4.7. *There exist $C_0 > 0$ and $\beta > 0$, independent of h and $\tilde{h}$, such that for each $h \le C_0 \tilde{h}$ there holds*

$$\sup_{\substack{v_h \in H_h \\ v_h \neq 0}} \frac{\langle \lambda_{\tilde{h}}, v_h \rangle}{\|v_h\|_{1,\Omega}} \ge \beta \|\lambda_{\tilde{h}}\|_{-1/2,\Gamma} \qquad \forall \lambda_{\tilde{h}} \in Q_{\tilde{h}}.$$

Proof. Given $\lambda_{\tilde{h}} \in Q_{\tilde{h}}$, we let $z \in H^1(\Omega)$ be the unique solution of the problem

$$-\Delta z + z = 0 \quad \text{in} \quad \Omega, \quad \nabla z \cdot \mathbf{n} = \lambda_{\tilde{h}} \quad \text{in} \quad \Gamma.$$

The continuous dependence result provided by the classical Lax–Milgram lemma (cf. Theorem 1.1) establishes that

$$\|z\|_{1,\Omega} \le c \|\lambda_{\tilde{h}}\|_{-1/2,\Gamma}.$$

In addition, since $Q_{\tilde{h}} \subseteq H^{\varepsilon}(\Gamma)$ for some $\varepsilon > 0$, it follows by elliptic regularity (cf. [42]) that $z \in H^{1+\delta}(\Omega) \quad \forall \delta \in [0, \delta_0]$, where $\delta_0 := \min\{\frac{1}{2} + \varepsilon, \frac{\pi}{\omega}\}$ and ω is the largest interior angle of Ω. We then fix $\delta \in (0, \delta_0)$, $\delta < 1/2$, and observe that

$$\|z\|_{1+\delta,\Omega} \le C \|\lambda_{\tilde{h}}\|_{-1/2+\delta,\Gamma}.$$

On the other hand, since [cf. (4.7)]

$$\|v - \mathbf{P}_{1,h}^1(v)\|_{1,\Omega} \le C h \|v\|_{2,\Omega} \qquad \forall v \in H^2(\Omega)$$

and clearly

$$\|v - \mathbf{P}_{1,h}^1(v)\|_{1,\Omega} \le h^0 \|v\|_{1,\Omega} \qquad \forall v \in H^1(\Omega),$$

the estimates for the interpolation of Sobolev spaces (cf. [49, Appendix B]) imply

$$\|v - \mathbf{P}_{1,h}^1(v)\|_{1,\Omega} \le C h^{\delta} \|v\|_{1+\delta,\Omega} \qquad \forall v \in H^{1+\delta}(\Omega).$$

It follows that

$$\|z - \mathbf{P}_{1,h}^1(z)\|_{1,\Omega} \le C h^{\delta} \|z\|_{1+\delta,\Omega} \le C h^{\delta} \|\lambda_{\tilde{h}}\|_{-1/2+\delta,\Gamma},$$

and using the inverse inequality for $Q_{\tilde{h}}$ (cf. Lemma 4.6), that is,

$$\|\lambda_{\tilde{h}}\|_{-1/2+\delta,\Gamma} \leq C\tilde{h}^{-\delta}\,\|\lambda_{\tilde{h}}\|_{-1/2,\Gamma},$$

we arrive at

$$\|z - \mathbf{P}^1_{1,h}(z)\|_{1,\Omega} \leq C\left\{\frac{h}{\tilde{h}}\right\}^{\delta}\|\lambda_{\tilde{h}}\|_{-1/2,\Gamma}. \tag{4.40}$$

Next, it is clear that

$$\|\mathbf{P}^1_{1,h}(z)\|_{1,\Omega} \leq \|z\|_{1,\Omega} \leq c\,\|\lambda_{\tilde{h}}\|_{-1/2,\Gamma}. \tag{4.41}$$

Then, using the Green identity in $H(\text{div};\Omega)$ (cf. Lemma 1.4) we see that

$$\begin{aligned}\langle\lambda_{\tilde{h}}, z\rangle &= \langle\nabla z\cdot\mathbf{n}, z\rangle = \langle\gamma_{\mathbf{n}}(\nabla z), \gamma_0(z)\rangle\\ &= \int_\Omega\left\{z\,\text{div}\,\nabla z + \nabla z\cdot\nabla z\right\} = \|z\|^2_{1,\Omega},\end{aligned}$$

and, recalling from Theorem 1.7 that $\gamma_{\mathbf{n}} : H(\text{div};\Omega) \to H^{-1/2}(\Gamma)$ is bounded, we obtain that

$$\|\lambda_{\tilde{h}}\|_{-1/2,\Gamma} = \|\gamma_{\mathbf{n}}(\nabla z)\|_{-1/2,\Gamma} \leq \|\nabla z\|_{\text{div},\Omega} = \|z\|_{1,\Omega},$$

which yields

$$\langle\lambda_{\tilde{h}}, z\rangle \geq \|\lambda_{\tilde{h}}\|^2_{-1/2,\Gamma}. \tag{4.42}$$

In this way, employing estimates (4.40)–(4.42) we find that

$$\begin{aligned}\sup_{\substack{v_h\in H_h\\ v_h\neq 0}}\frac{\langle\lambda_{\tilde{h}}, v_h\rangle}{\|v_h\|_{1,\Omega}} &\geq \frac{|\langle\lambda_{\tilde{h}}, \mathbf{P}^1_{1,h}(z)\rangle|}{\|\mathbf{P}^1_{1,h}(z)\|_{1,\Omega}}\\ &\geq \frac{|\langle\lambda_{\tilde{h}}, \mathbf{P}^1_{1,h}(z)\rangle|}{c\,\|\lambda_{\tilde{h}}\|_{-1/2,\Gamma}}\\ &\geq \tilde{C}\left\{\frac{|\langle\lambda_{\tilde{h}}, z\rangle|}{\|\lambda_{\tilde{h}}\|_{-1/2,\Gamma}} - \frac{|\langle\lambda_{\tilde{h}}, z - \mathbf{P}^1_{1,h}(z)\rangle|}{\|\lambda_{\tilde{h}}\|_{-1/2,\Gamma}}\right\}\\ &\geq \tilde{C}\|\lambda_{\tilde{h}}\|_{-1/2,\Gamma} - C\left\{\frac{h}{\tilde{h}}\right\}^{\delta}\|\lambda_{\tilde{h}}\|_{-1/2,\Gamma}\\ &= \left\{\tilde{C} - C\left(\frac{h}{\tilde{h}}\right)^{\delta}\right\}\|\lambda_{\tilde{h}}\|_{-1/2,\Gamma},\end{aligned}$$

where, choosing $h \leq C_0\,\tilde{h}$, with $C_0 = \left(\frac{\tilde{C}}{2C}\right)^{1/\delta}$, we deduce the existence of $\beta > 0$ such that

$$\sup_{\substack{v_h \in H_h \\ v_h \neq 0}} \frac{\langle \lambda_{\tilde{h}}, v_h \rangle}{\|v_h\|_{1,\Omega}} \geq \beta \, \|\lambda_{\tilde{h}}\|_{-1/2,\Gamma} \qquad \forall \lambda_{\tilde{h}} \in Q_{\tilde{h}},$$

thereby completing the proof. □

Consequently, applying the results from the discrete Babuška–Brezzi theory (cf. Theorems 2.4 and 2.6), we deduce that $\forall h \leq C_0 \tilde{h}$ there exists a unique pair $(u_h, \xi_{\tilde{h}}) \in H_h \times Q_{\tilde{h}}$ solution of the Galerkin scheme (4.37), and there holds the Cea estimate

$$\begin{aligned} &\|u - u_h\|_{1,\Omega} + \|\xi - \xi_{\tilde{h}}\|_{-1/2,\Gamma} \\ &\quad \leq c \left\{ \inf_{v_h \in H_h} \|u - v_h\|_{1,\Omega} + \inf_{\lambda_{\tilde{h}} \in Q_{\tilde{h}}} \|\xi - \lambda_{\tilde{h}}\|_{-1/2,\Gamma} \right\}. \end{aligned} \tag{4.43}$$

Note that the first term on the right-hand side of the preceding equation reduces to $\|u - \mathbf{P}^1_{1,h}(u)\|_{1,\Omega}$, which can be bounded by means of (4.7). To estimate the second term we need the following previous result.

Lemma 4.8. *Let $\mathscr{P}^0_{\tilde{h}} : L^2(\Gamma) \to Q_{\tilde{h}}$ be the orthogonal projector with respect to the $L^2(\Gamma)$-inner product. Then there holds*

$$\|\lambda - \mathscr{P}^0_{\tilde{h}}(\lambda)\|_{-1/2,\Gamma} \leq c \tilde{h} \, \|\lambda\|_{1/2,\Gamma} \qquad \forall \lambda \in H^{1/2}(\Gamma).$$

Proof. Starting from the estimates

$$\|\lambda - \mathscr{P}^0_{\tilde{h}}(\lambda)\|_{0,\Gamma} \leq \tilde{h}^0 \, \|\lambda\|_{0,\Gamma} \quad \forall \lambda \in L^2(\Gamma)$$

and

$$\|\lambda - \mathscr{P}^0_{\tilde{h}}(\lambda)\|_{0,\Gamma} \leq C \tilde{h} \, \|\lambda\|_{1,\Gamma} \quad \forall \lambda \in H^1(\Gamma),$$

the latter being a consequence of the Deny–Lions and Bramble–Hilbert Lemmas (cf. Theorems 3.4 and 3.5), we find by interpolation that

$$\|\lambda - \mathscr{P}^0_{\tilde{h}}(\lambda)\|_{0,\Gamma} \leq C \tilde{h}^{1/2} \, \|\lambda\|_{1/2,\Gamma} \qquad \forall \lambda \in H^{1/2}(\Gamma).$$

Next, using a duality argument and the preceding estimate, we have that for each $\lambda \in H^{1/2}(\Gamma)$ there holds

$$\begin{aligned} \|\lambda - \mathscr{P}^0_{\tilde{h}}(\lambda)\|_{-1/2,\Gamma} &= \sup_{\substack{\eta \in H^{1/2}(\Gamma) \\ \eta \neq \Theta}} \frac{\langle \lambda - \mathscr{P}^0_{\tilde{h}}(\lambda), \eta \rangle}{\|\eta\|_{1/2,\Gamma}} \\ &= \sup_{\substack{\eta \in H^{1/2}(\Gamma) \\ \eta \neq \Theta}} \frac{\langle \lambda - \mathscr{P}^0_{\tilde{h}}(\lambda), \eta \rangle_{0,\Gamma}}{\|\eta\|_{1/2,\Gamma}} \end{aligned}$$

$$
\begin{aligned}
&= \sup_{\substack{\eta \in H^{1/2}(\Gamma)\\ \eta \neq 0}} \frac{\langle \lambda - \mathscr{P}^0_{\tilde{h}}(\lambda), \eta - \mathscr{P}^0_{\tilde{h}}(\eta)\rangle_{0,\Gamma}}{\|\eta\|_{1/2,\Gamma}} \\
&\leq \sup_{\substack{\eta \in H^{1/2}(\Gamma)\\ \eta \neq 0}} \frac{\|\lambda - \mathscr{P}^0_{\tilde{h}}(\lambda)\|_{0,\Gamma}\, \|\eta - \mathscr{P}^0_{\tilde{h}}(\eta)\|_{0,\Gamma}}{\|\eta\|_{1/2,\Gamma}} \\
&\leq \sup_{\substack{\eta \in H^{1/2}(\Gamma)\\ \eta \neq 0}} \frac{C\tilde{h}^{1/2}\|\lambda\|_{1/2,\Gamma}\, C\tilde{h}^{1/2}\|\eta\|_{1/2,\Gamma}}{\|\eta\|_{1/2,\Gamma}} \\
&= \tilde{C}\tilde{h}\|\lambda\|_{1/2,\Gamma},
\end{aligned}
$$

which completes the proof. □

We now let $\mathscr{P}_{-1/2,\tilde{h}} : H^{-1/2}(\Gamma) \to Q_{\tilde{h}}$ be the orthogonal projector with respect to the $H^{-1/2}(\Gamma)$-inner product. It is then clear that

$$
\inf_{\lambda_{\tilde{h}} \in Q_{\tilde{h}}} \|\xi - \lambda_{\tilde{h}}\|_{-1/2,\Gamma} = \|\xi - \mathscr{P}_{-1/2,\tilde{h}}(\xi)\|_{-1/2,\Gamma}
$$

and that

$$
\|\xi - \mathscr{P}_{-1/2,\tilde{h}}(\xi)\|_{-1/2,\Gamma} \leq \|\xi\|_{-1/2,\Gamma} \qquad \forall \xi \in H^{-1/2}(\Gamma). \tag{4.44}
$$

In addition, utilizing Lemma 4.8 we obtain that

$$
\|\xi - \mathscr{P}_{-1/2,\tilde{h}}(\xi)\|_{-1/2,\Gamma} \leq \|\xi - \mathscr{P}^0_{\tilde{h}}(\xi)\|_{-1/2,\Gamma} \leq C\tilde{h}\|\xi\|_{1/2,\Gamma} \quad \forall \xi \in H^{1/2}(\Gamma),
$$

that is,

$$
\|\xi - \mathscr{P}_{-1/2,\tilde{h}}(\xi)\|_{-1/2,\Gamma} \leq C\tilde{h}\|\xi\|_{1/2,\Gamma} \qquad \forall \xi \in H^{1/2}(\Gamma),
$$

which, together with (4.44) and thanks to the interpolation estimates for Sobolev spaces, gives

$$
\|\xi - \mathscr{P}_{-1/2,\tilde{h}}(\xi)\|_{-1/2,\Gamma} \leq C\tilde{h}^{r+\frac{1}{2}}\|\xi\|_{r,\Gamma} \quad \forall \xi \in H^r(\Gamma), \forall r \in [-1/2, 1/2]. \tag{4.45}
$$

Therefore, recalling that

$$
\|v - \mathbf{P}^1_{1,h}(v)\|_{1,\Omega} \leq Ch^l\|v\|_{l+1,\Omega} \quad \forall v \in H^{l+1}(\Omega),\ 0 \leq l \leq 1, \tag{4.46}
$$

we conclude from (4.43), (4.45), and (4.46) that

$$
\|u - u_h\|_{1,\Omega} + \|\xi - \xi_{\tilde{h}}\|_{-1/2,\Gamma} \leq C\left\{h^l\|u\|_{l+1,\Omega} + \tilde{h}^{r+1/2}\|\xi\|_{r,\Gamma}\right\}
$$

for each $u \in H^{l+1}(\Omega)$, $0 \leq l \leq 1$, and for each $\xi \in H^r(\Gamma)$, $-1/2 \leq r \leq 1/2$.

4.4 Poisson Problem with Neumann Boundary Conditions

In this section we analyze a Galerkin scheme for the two-dimensional version of the Poisson problem studied in Sect. 2.4.2 with Neumann boundary conditions, that is, when $\Gamma_N = \Gamma$. In other words, given Ω a bounded domain of $\mathbb{R}^2$, $f \in L^2(\Omega)$, and $g \in H^{-1/2}(\Gamma)$, we are interested in the boundary value problem

$$-\Delta u = f \quad \text{in} \quad \Omega, \qquad \frac{\partial u}{\partial \mathbf{n}} = g \quad \text{in} \quad \Gamma, \qquad \int_\Omega u = 0,$$

for which we need to assume that the data satisfy the compatibility condition

$$\int_\Omega f + \langle g, 1\rangle = 0.$$

Then defining the auxiliary unknowns

$$\sigma := \nabla u \quad \text{in} \quad \Omega \quad \text{and} \quad \xi := -\gamma_0(u) \quad \text{in} \quad \Gamma$$

and proceeding as in Sects. 2.4.1 and 2.4.2, one arrives at the mixed variational formulation: find $(\sigma, (u, \xi)) \in H \times Q$ such that

$$\begin{aligned} a(\sigma, \tau) + b(\tau, (u, \xi)) &= 0 \qquad \forall \tau \in H, \\ b(\sigma, (v, \lambda)) \qquad\qquad &= -\int_\Omega f v + \langle g, \lambda\rangle \qquad \forall (v, \lambda) \in Q, \end{aligned}$$

where $H := H(\mathrm{div}; \Omega)$, $Q := L_0^2(\Omega) \times H^{1/2}(\Gamma)$, and the bounded bilinear forms $a : H \times H \to \mathbb{R}$ and $b : H \times Q \to \mathbb{R}$ are defined by

$$a(\sigma, \tau) := \int_\Omega \sigma \cdot \tau \quad \forall \sigma, \tau \in H \tag{4.47}$$

and

$$b(\tau, (v, \lambda)) := \int_\Omega v \,\mathrm{div}\, \tau + \langle \tau \cdot \mathbf{n}, \lambda\rangle \quad \forall \tau \in H, \forall (v, \lambda) \in Q. \tag{4.48}$$

As in Chap. 3, henceforth we omit the symbol $\gamma_{\mathbf{n}}$ to denote the respective normal traces and simply write $\tau \cdot \mathbf{n}$ instead of $\gamma_{\mathbf{n}}(\tau)$.

We now consider finite-dimensional subspaces $H_h \subseteq H$, $Q_h^u \subseteq L_0^2(\Omega)$ and $Q_h^\xi \subseteq H^{1/2}(\Gamma)$ and define

$$Q_h := Q_h^u \times Q_h^\xi \subseteq Q.$$

Then, the associated Galerkin scheme reduces to the following formulation: find $(\sigma_h, (u_h, \xi_h)) \in H_h \times Q_h$ such that

$$\begin{aligned} a(\sigma_h, \tau_h) + b(\tau_h, (u_h, \xi_h)) &= 0 \quad \forall \tau_h \in H_h, \\ b(\sigma_h, (v_h, \lambda_h)) \qquad\qquad &= -\int_\Omega f v_h + \langle g, \lambda_h\rangle \quad \forall (v_h, \lambda_h) \in Q_h. \end{aligned} \tag{4.49}$$

For the analysis of (4.49) we first focus on the discrete inf-sup condition for b, that is, on the eventual existence of $\beta > 0$, independent of h, such that

$$\sup_{\substack{\tau_h \in H_h \\ \tau_h \neq 0}} \frac{b(\tau_h,(v_h,\lambda_h))}{\|\tau_h\|_H} \geq \beta\, \|(v_h,\lambda_h)\|_Q \quad \forall\, (v_h,\lambda_h) \in Q_h. \tag{4.50}$$

Since b can be decomposed as the sum of two bilinear forms b_1 and b_2, that is [cf. (4.48)],

$$b(\tau,(v,\lambda)) = b_1(\tau,v) + b_2(\tau,\lambda) \quad \forall\, \tau \in H, \forall\, (v,\lambda) \in Q, \tag{4.51}$$

we could certainly utilize the corresponding characterization result established in [40, Theorem 7] to prove (4.50). Alternatively, and due to the same decomposition, one could also employ the slightly different equivalence given in [45, Theorem 3.1]. However, and to provide further points of view to this analysis, in what follows we apply another procedure that can be seen as a combination of the aforementioned approaches. Indeed, we first use the boundedness of the normal trace of vectors in $H(\mathrm{div};\Omega)$ [see (1.44) in the proof of Theorem 1.7] to deduce that

$$\begin{aligned} \sup_{\substack{\tau_h \in H_h \\ \tau_h \neq 0}} \frac{b(\tau_h,(v_h,\lambda_h))}{\|\tau_h\|_H} &\geq \sup_{\substack{\tau_h \in H_h \\ \tau_h \neq 0}} \frac{b_1(\tau_h,v_h)}{\|\tau_h\|_{\mathrm{div},\Omega}} - \|\lambda_h\|_{1/2,\Gamma} \\ &= \sup_{\substack{\tau_h \in H_h \\ \tau_h \neq 0}} \frac{\displaystyle\int_\Omega v_h \,\mathrm{div}\, \tau_h}{\|\tau_h\|_{\mathrm{div},\Omega}} - \|\lambda_h\|_{1/2,\Gamma} \quad \forall\, (v_h,\lambda_h) \in Q_h. \end{aligned}$$

Next, considering the particular subspaces [cf. (4.1) and (4.3)]

$$H_h := H_h^0 := \Big\{ \tau_h \in H(\mathrm{div};\Omega) : \quad \tau_h|_K \in RT_0(K) \quad \forall\, K \in \mathcal{T}_h \Big\},$$

$$Q_h^u := Y_h^0 \cap L_0^2(\Omega) := \Big\{ v_h \in L_0^2(\Omega) : \quad v_h|_K \in \mathbb{P}_0(K) \quad \forall\, K \in \mathcal{T}_h \Big\}$$

and employing the analysis from Sect. 4.2, we obtain that

$$\sup_{\substack{\tau_h \in H_h \\ \tau_h \neq 0}} \frac{b(\tau_h,(v_h,\lambda_h))}{\|\tau_h\|_H} \geq \hat{\beta}\, \|v_h\|_{0,\Omega} - \|\lambda_h\|_{1/2,\Gamma} \quad \forall\, (v_h,\lambda_h) \in Q_h, \tag{4.52}$$

with a constant $\hat{\beta} > 0$, independent of h. Then, it is straightforward to see that

$$\begin{aligned} \sup_{\substack{\tau_h \in H_h \\ \tau_h \neq 0}} \frac{b(\tau_h,(v_h,\lambda_h))}{\|\tau_h\|_H} &\geq \sup_{\substack{\tau_h \in V_{1,h} \\ \tau_h \neq 0}} \frac{b_2(\tau_h,\lambda_h)}{\|\tau_h\|_{\mathrm{div},\Omega}} \\ &= \sup_{\substack{\tau_h \in V_{1,h} \\ \tau_h \neq 0}} \frac{\langle \tau_h \cdot \mathbf{n}, \lambda_h \rangle}{\|\tau_h\|_{\mathrm{div},\Omega}} \qquad \forall\, (v_h,\lambda_h) \in Q_h, \end{aligned} \tag{4.53}$$

where $V_{1,h}$ is the discrete kernel of b_1, that is,

$$\begin{aligned} V_{1,h} &= \left\{ \tau_h \in H_h : \quad b_1(\tau_h, v_h) := \int_\Omega v_h \operatorname{div} \tau_h = 0 \quad \forall v_h \in Q_h^u \right\} \\ &= \left\{ \tau_h \in H_h : \quad \operatorname{div} \tau_h \in \mathbb{P}_0(\Omega) \right\}. \end{aligned} \tag{4.54}$$

Hence, it is easy to see from (4.52) and (4.53) that, in order to conclude (4.50), it suffices to show that there exists a constant $\tilde{\beta} > 0$, independent of h, such that

$$\sup_{\substack{\tau_h \in V_{1,h} \\ \tau_h \neq 0}} \frac{\langle \tau_h \cdot \mathbf{n}, \lambda_h \rangle}{\|\tau_h\|_{\operatorname{div},\Omega}} \geq \tilde{\beta} \, \|\lambda_h\|_{1/2,\Gamma} \qquad \forall \lambda_h \in Q_h^\xi . \tag{4.55}$$

Throughout the rest of this section we aim to prove (4.55). To this end, we now introduce the following definition.

Definition 4.1. Let $\Phi_h(\Gamma) := \{\tau_h \cdot \mathbf{n}|_\Gamma : \quad \tau_h \in V_{1,h}\}$. We say that a linear operator $\mathscr{L}_h : \Phi_h(\Gamma) \to V_{1,h}$ is a STABLE DISCRETE LIFTING if

(i) $\mathscr{L}_h(\phi_h) \cdot \mathbf{n} = \phi_h \quad \text{on} \quad \Gamma \quad \forall \phi_h \in \Phi_h(\Gamma)$;
(ii) $\exists c > 0$, independently of h such that

$$\|\mathscr{L}_h(\phi_h)\|_{\operatorname{div},\Omega} \leq c \, \|\phi_h\|_{-1/2,\Gamma} \qquad \forall \phi_h \in \Phi_h(\Gamma).$$

Lemma 4.9. *Assume that there exists a stable discrete lifting $\mathscr{L}_h : \Phi_h(\Gamma) \to V_{1,h}$. Then the discrete inf-sup condition* (4.55) *is equivalent to the existence of $C > 0$, independent of h, such that*

$$\sup_{\substack{\phi_h \in \Phi_h(\Gamma) \\ \phi_h \neq 0}} \frac{\langle \phi_h, \lambda_h \rangle}{\|\phi_h\|_{-1/2,\Gamma}} \geq C \|\lambda_h\|_{1/2,\Gamma} \qquad \forall \lambda_h \in Q_h^\xi . \tag{4.56}$$

Proof. It suffices to see, according to (ii) in Definition 4.1, that

$$\frac{|\langle \phi_h, \lambda_h \rangle|}{\|\phi_h\|_{-1/2,\Gamma}} \leq \frac{c \, |\langle \phi_h, \lambda_h \rangle|}{\|\mathscr{L}_h(\phi_h)\|_{\operatorname{div},\Omega}} \leq c \sup_{\substack{\tau_h \in V_{1,h} \\ \tau_h \neq 0}} \frac{|\langle \tau_h \cdot \mathbf{n}, \lambda_h \rangle|}{\|\tau_h\|_{\operatorname{div},\Omega}}$$

and, according to the bound given by (1.44) (cf. Theorem 1.7), that

$$\frac{|\langle \tau_h \cdot \mathbf{n}, \lambda_h \rangle|}{\|\tau_h\|_{\operatorname{div},\Omega}} \leq \frac{|\langle \tau_h \cdot \mathbf{n}, \lambda_h \rangle|}{\|\tau_h \cdot \mathbf{n}\|_{-1/2,\Gamma}} \leq \sup_{\substack{\phi_h \in \Phi_h(\Gamma) \\ \phi_h \neq 0}} \frac{|\langle \phi_h, \lambda_h \rangle|}{\|\phi_h\|_{-1/2,\Gamma}} ,$$

whence (4.55) and (4.56) are equivalent. □

In what follows we provide sufficient conditions for the existence of a stable discrete lifting $\mathscr{L}_h : \Phi_h(\Gamma) \to V_{1,h}$. To this end, we proceed as in [39, Sects. 5.2

and 5.3] and assume that $\mathscr{T}_h$ is quasi-uniform around Γ. This means that there exist a neighborhood Ω_Γ of Γ and a constant $c > 0$, independent of h, such that, denoting

$$\mathscr{T}_{h,\Gamma} := \{K \in \mathscr{T}_h : \quad K \cap \Omega_\Gamma \neq \emptyset\}, \tag{4.57}$$

there holds

$$\max_{K \in \mathscr{T}_{h,\Gamma}} h_K \leq c \min_{K \in \mathscr{T}_{h,\Gamma}} h_K.$$

It is important to remark here that, while the aforementioned requirement of quasi-uniformity was removed recently in [48, Sects. 4 and 5] for the two-dimensional case, we prefer to keep it throughout the rest of the present analysis since, to our knowledge, the approach to be shown below is also the only known one that can be applied to derive the existence of stable discrete liftings in three dimensions (e.g., [33, Lemma 7.5]).

Now, because of the regularity of $\mathscr{T}_h$ [cf. (3.32)], which means that

$$\frac{h_K}{\rho_K} \leq c \quad \forall K \in \mathscr{T}_h, \quad \forall h > 0$$

or, equivalently, that $\{\mathscr{T}_h\}_{h>0}$ satisfies the minimum angle condition, the quasi-uniformity assumption implies that the partition on Γ inherited from $\mathscr{T}_h$, say Γ_h, is also quasi-uniform, that is, there exists $c > 0$, independently of h, such that

$$h_\Gamma := \max\Big\{|e| : \quad e \in \Gamma_h\Big\} \leq c \min\Big\{|e| : \quad e \in \Gamma_h\Big\}.$$

We now define

$$\tilde{\Phi}_h(\Gamma) := \Big\{\phi_h \in L^2(\Gamma) : \quad \phi_h|_e \in \mathbb{P}_0(e) \qquad \forall e \in \Gamma_h\Big\}$$

and notice that $\Phi_h(\Gamma) \subseteq \tilde{\Phi}_h(\Gamma)$. In addition, the quasi-uniformity of Γ_h implies that $\tilde{\Phi}_h(\Gamma)$, which coincides with the space $Q_{\tilde{h}}$ given in Sect. 4.3, satisfies the inverse inequality (cf. Lemma 4.6)

$$\|\phi_h\|_{-1/2+\delta,\Gamma} \leq C h_\Gamma^{-\delta} \|\phi_h\|_{-1/2,\Gamma} \qquad \forall \phi_h \in \tilde{\Phi}_h(\Gamma), \qquad \forall \delta \in [0,1/2]. \tag{4.58}$$

Theorem 4.1. *Under the previously stated assumptions, there exists a stable discrete lifting $\mathscr{L}_h : \Phi_h(\Gamma) \to V_{1,h}$.*

Proof. Let $\phi_h \in \tilde{\Phi}_h(\Gamma)$, and let $v \in H^1(\Omega)$ be the unique solution of the problem

$$\Delta v = \frac{1}{|\Omega|}\int_\Gamma \phi_h \quad \text{in} \quad \Omega, \quad \nabla v \cdot \mathbf{n} = \phi_h \quad \text{on} \quad \Gamma, \quad \int_\Omega v = 0.$$

The corresponding continuous dependence result says that $\|v\|_{1,\Omega} \leq C_1 \|\phi_h\|_{-1/2,\Gamma}$. Next, the elliptic regularity result in nonconvex polygonal domains (cf. [42]) establishes that there exists $\delta \in (0,1/2)$ such that $v \in H^{1+\delta}(\Omega)$ and

$$\|v\|_{1+\delta,\Omega} \leq C \|\phi_h\|_{-1/2+\delta,\Gamma}\,.$$

It follows that $\nabla v \in [H^{\delta}(\Omega)]^2 \cap H(\mathrm{div};\Omega)$ [note that $\mathrm{div}(\nabla v) = \Delta v = \frac{1}{|\Omega|}\int_\Gamma \phi_h \in \mathbb{R}$], and hence we can define (see remark immediately preceding Lemma 3.19)

$$\mathscr{L}_h(\phi_h) := \Pi_h^0(\nabla v)\,.$$

According to the preceding equation and (3.8) (cf. Lemma 3.7), we obtain

$$\mathrm{div}\,\mathscr{L}_h(\phi_h) = \mathrm{div}\,\Pi_h^0(\nabla v) = \mathscr{P}_h^0(\mathrm{div}\,\nabla v) = \mathscr{P}_h^0(\Delta v) = \frac{1}{|\Omega|}\int_\Gamma \phi_h \in \mathbb{P}_0(\Omega)\,,$$

which confirms that $\mathscr{L}_h(\phi_h) \in V_{1,h}$, and, thanks to (3.36) (see proof of Lemma 3.18), we find that

$$\mathscr{L}_h(\phi_h)\cdot\mathbf{n} = \Pi_h^0(\nabla v)\cdot\mathbf{n} = \mathscr{P}_{h,\Gamma}^0(\nabla v\cdot\mathbf{n}) = \mathscr{P}_{h,\Gamma}^0(\phi_h) = \phi_h,$$

where $\mathscr{P}_{h,\Gamma}^0 : L^2(\Gamma) \longrightarrow \tilde{\Phi}_h(\Gamma)$ is the orthogonal projector. Note that this last identity also proves that $\tilde{\Phi}_h(\Gamma) \subseteq \Phi_h(\Gamma)$, and therefore we deduce that

$$\begin{aligned}\Phi_h(\Gamma) &:= \Big\{\tau_h\cdot\mathbf{n}|_\Gamma : \quad \tau_h \in V_{1,h}\Big\}\\ &= \tilde{\Phi}_h(\Gamma) := \Big\{\phi_h \in L^2(\Gamma) : \quad \phi_h|_e \in \mathbb{P}_0(e) \quad \forall e \in \Gamma_h\Big\}.\end{aligned}$$

It remains to prove that $\mathscr{L}_h : \Phi_h(\Gamma) \to V_{1,h}$ is uniformly bounded. To this end, we first observe that

$$\|\mathscr{L}_h(\phi_h)\|_{\mathrm{div},\Omega}^2 = \|\mathscr{L}_h(\phi_h)\|_{0,\Omega}^2 + \left\|\frac{1}{|\Omega|}\int_\Gamma \phi_h\right\|_{0,\Omega}^2 \leq \|\mathscr{L}_h(\phi_h)\|_{0,\Omega}^2 + C\,\|\phi_h\|_{-1/2,\Gamma}^2\,.$$

Next, we recall from (4.57) the definition of $\mathscr{T}_{h,\Gamma}$ and introduce the sets

$$\Omega_h^1 := \cup\Big\{K \in \mathscr{T}_h : \quad K \notin \mathscr{T}_{h,\Gamma}\Big\} \subseteq \Omega\backslash\Omega_\Gamma$$

and

$$\Omega_h^2 := \Omega\backslash\Omega_h^1 = \cup\Big\{K \in \mathscr{T}_{h,\Gamma}\Big\}.$$

Since $\Omega\backslash\Omega_\Gamma$ is strictly contained in Ω, the interior elliptic regularity result (cf. [49, Theorem 4.16]) implies that $v|_{\Omega\backslash\Omega_\Gamma} \in H^2(\Omega\backslash\Omega_\Gamma)$ and

$$\|v\|_{2,\Omega\backslash\Omega_\Gamma} \leq C_2\,\|\phi_h\|_{-1/2,\Gamma}\,.$$

It follows that

$$
\begin{aligned}
\|\mathscr{L}_h(\phi_h)\|_{0,\Omega} &\leq \|\mathscr{L}_h(\phi_h)\|_{0,\Omega_h^1} + \|\mathscr{L}_h(\phi_h)\|_{0,\Omega_h^2} \\
&= \|\Pi_h^0(\nabla v)\|_{0,\Omega_h^1} + \|\Pi_h^0(\nabla v)\|_{0,\Omega_h^2} \\
&\leq C\|\nabla v\|_{1,\Omega_h^1} + \|\nabla v\|_{0,\Omega_h^2} + \|\nabla v - \Pi_h^0(\nabla v)\|_{0,\Omega_h^2} \\
&\leq C\|v\|_{2,\Omega_h^1} + \|v\|_{1,\Omega_h^2} + \|\nabla v - \Pi_h^0(\nabla v)\|_{0,\Omega_h^2} \\
&\leq CC_2\,\|\phi_h\|_{-1/2,\Gamma} + C_1\,\|\phi_h\|_{-1/2,\Gamma} + \|\nabla v - \Pi_h^0(\nabla v)\|_{0,\Omega_h^2}.
\end{aligned}
$$

On the other hand, applying estimate (3.37) (cf. Lemma 3.19) and inverse inequality (4.58), we obtain

$$
\begin{aligned}
\|\nabla v - \Pi_h^0(\nabla v)\|_{0,\Omega_h^2}^2 &= \sum_{K\in\mathscr{T}_{h,\Gamma}} \|\nabla v - \Pi_K^0(\nabla v)\|_{0,K}^2 \\
&\leq C \sum_{K\in\mathscr{T}_{h,\Gamma}} h_K^{2\delta}\left\{|\nabla v|_{\delta,K}^2 + \left\|\frac{1}{|\Omega|}\int_\Gamma \phi_h\right\|_{0,K}^2\right\} \\
&\leq \overline{C} \max_{K\in\mathscr{T}_{h,\Gamma}} h_K^{2\delta}\left\{\|v\|_{1+\delta,\Omega_h^2}^2 + \|\phi_h\|_{-1/2,\Gamma}^2\right\} \\
&\leq \overline{C} \max_{K\in\mathscr{T}_{h,\Gamma}} h_K^{2\delta}\left\{\|v\|_{1+\delta,\Omega}^2 + \|\phi_h\|_{-1/2,\Gamma}^2\right\} \\
&\leq C \max_{K\in\mathscr{T}_{h,\Gamma}} h_K^{2\delta}\left\{\|\phi_h\|_{-1/2+\delta,\Gamma}^2 + \|\phi_h\|_{-1/2,\Gamma}^2\right\} \\
&\leq C \max_{K\in\mathscr{T}_{h,\Gamma}} h_K^{2\delta}\left\{h_\Gamma^{-2\delta}\,\|\phi_h\|_{-1/2,\Gamma}^2 + \|\phi_h\|_{-1/2,\Gamma}^2\right\} \\
&\leq C\|\phi_h\|_{-1/2,\Gamma}^2,
\end{aligned}
$$

where the fact that $h_K \leq C h_\Gamma \quad \forall K \in \mathscr{T}_{h,\Gamma}$ has been used in the last inequality. Consequently, gathering together the preceding estimates, we conclude that

$$
\|\mathscr{L}_h(\phi_h)\|_{\mathrm{div},\Omega} \leq C\|\phi_h\|_{-1/2,\Gamma} \qquad \forall \phi_h \in \Phi_h(\Gamma),
$$

which completes the proof. □

We now aim to prove the following result, which, according to Lemma 4.9, will suffice to conclude the required discrete inf-sup condition for the term on Γ.

Lemma 4.10. *There exists $\beta > 0$, independent of h, such that*

$$
\sup_{\substack{\phi_h\in\Phi_h(\Gamma)\\ \phi_h\neq 0}} \frac{\langle \phi_h, \lambda_h\rangle}{\|\phi_h\|_{-1/2,\Gamma}} \geq \beta\,\|\lambda_h\|_{1/2,\Gamma} \qquad \forall \lambda_h \in Q_h^\xi.
$$

For the aforementioned purpose we proceed as in [39, Sect. 5.3] and describe in what follows two different procedures under which Lemma 4.10 is proved.

PROCEDURE 1. The subsequent analysis is based on the approach originally proposed in [9]. In fact, we now set

$$\Phi_h(\Gamma) = \tilde{\Phi}_h(\Gamma) := \left\{ \phi_h \in L^2(\Gamma) : \quad \phi_h|_e \in \mathbb{P}_0(e) \quad \forall e \in \Gamma_h \right\},$$

where Γ_h is the partition on Γ inherited from $\mathscr{T}_h$, and let $h_\Gamma := \max \left\{ |e| : e \in \Gamma_h \right\}$. In addition, we introduce the space

$$Q_{\tilde{h}}^{\xi} := \left\{ \lambda_{\tilde{h}} \in C(\Gamma) : \quad \lambda_{\tilde{h}}|_{\tilde{\Gamma}_j} \in \mathbb{P}_1(\tilde{\Gamma}_j) \qquad \forall j \in \{1, \cdots, m\} \right\},$$

where $\left\{ \tilde{\Gamma}_1, \tilde{\Gamma}_2, \cdots, \tilde{\Gamma}_m \right\}$ is another partition of Γ and

$$\tilde{h} := \max \left\{ |\tilde{\Gamma}_j| : j \in \{1, \cdots, m\} \right\}.$$

Then we have the following result.

Lemma 4.11. *There exist c_0, $\beta > 0$, independent of h_Γ and $\tilde{h}$, such that $\forall h_\Gamma \leq c_0 \tilde{h}$*

$$\sup_{\substack{\phi_h \in \Phi_h(\Gamma) \\ \phi_h \neq 0}} \frac{\langle \phi_h, \lambda_{\tilde{h}} \rangle}{\|\phi_h\|_{-1/2,\Gamma}} \geq \beta \, \|\lambda_{\tilde{h}}\|_{1/2,\Gamma} \qquad \forall \lambda_{\tilde{h}} \in Q_{\tilde{h}}^{\xi}.$$

Proof. Given $\lambda_{\tilde{h}} \in Q_{\tilde{h}}^{\xi}$, we let $z \in H^1(\Omega)$ be the unique solution of the problem

$$-\Delta z + z = 0 \quad \text{in} \quad \Omega, \quad z = \lambda_{\tilde{h}} \quad \text{on} \quad \Gamma.$$

Since $Q_{\tilde{h}}^{\xi} \subseteq H^1(\Gamma)$, we obtain by elliptic regularity (cf. [42]) that $z \in H^{1+\delta}(\Omega)$ and $\|z\|_{1+\delta,\Omega} \leq C \|\lambda_{\tilde{h}}\|_{1/2+\delta,\Gamma}$ $\forall \delta \in [0, \delta_0]$, where $\delta_0 := \min \left\{ \frac{1}{2}, \frac{\pi}{\omega} \right\}$ and ω is the largest interior angle of Ω. In what follows, we fix $\delta \in (0, \delta_0]$, $\delta < 1/2$, and observe that $\nabla z \cdot \mathbf{n}\Big|_\Gamma \in H^{-1/2+\delta}(\Gamma)$ and

$$\left\| \nabla z \cdot \mathbf{n} \right\|_{-1/2+\delta,\Gamma} \leq C \|z\|_{1+\delta,\Omega} \, .$$

Then, according to the approximation properties of $\Phi_h(\Gamma)$, whose details are described following this proof, we find that

$$\begin{aligned} \left\| \nabla z \cdot \mathbf{n} - \mathscr{P}_h^{-1/2}(\nabla z \cdot \mathbf{n}) \right\|_{-1/2,\Gamma} &\leq C h_\Gamma^{\delta} \left\| \nabla z \cdot \mathbf{n} \right\|_{-1/2+\delta,\Gamma} \\ &\leq C h_\Gamma^{\delta} \, \|z\|_{1+\delta,\Omega} \leq C h_\Gamma^{\delta} \, \|\lambda_{\tilde{h}}\|_{1/2+\delta,\Gamma}, \end{aligned}$$

where $\mathscr{P}_h^{-1/2} : H^{-1/2}(\Gamma) \to \Phi_h(\Gamma)$ is the orthogonal projector with respect to the $H^{-1/2}(\Gamma)$-inner product. Hence, applying the inverse inequality for $Q_{\tilde{h}}^{\xi}$ (see remark following proof), we deduce that

$$\left\|\nabla z\cdot\mathbf{n}-\mathscr{P}_h^{-1/2}(\nabla z\cdot\mathbf{n})\right\|_{-1/2,\Gamma}\le C\left(\frac{h_\Gamma}{\tilde{h}}\right)^{\delta}\|\lambda_{\tilde{h}}\|_{1/2,\Gamma}.$$

Next, using that $\|\nabla z\|_{\mathrm{div},\Omega}=\|z\|_{1,\Omega}$ and that $\Delta z=z$ in Ω, it follows that

$$\left\|\mathscr{P}_h^{-1/2}(\nabla z\cdot\mathbf{n})\right\|_{-1/2,\Gamma}\le\left\|\nabla z\cdot\mathbf{n}\right\|_{-1/2,\Gamma}\le\|z\|_{1,\Omega}\le\overline{C}\,\|\lambda_{\tilde{h}}\|_{1/2,\Gamma}.$$

On the other hand, it is clear that

$$\left\langle\nabla z\cdot\mathbf{n},\lambda_{\tilde{h}}\right\rangle=\left\langle\nabla z\cdot\mathbf{n},z\right\rangle=\|z\|_{1,\Omega}^2\ge\tilde{C}\,\|z\|_{1/2,\Gamma}^2=\hat{C}\,\|\lambda_{\tilde{h}}\|_{1/2,\Gamma}^2.$$

In this way, we deduce that

$$\begin{aligned}\sup_{\substack{\phi_h\in\Phi_h(\Gamma)\\ \phi_h\neq0}}\frac{\langle\phi_h,\lambda_{\tilde{h}}\rangle}{\|\phi_h\|_{-1/2,\Gamma}}&\ge\frac{\left\langle\mathscr{P}_h^{-1/2}(\nabla z\cdot\mathbf{n}),\lambda_{\tilde{h}}\right\rangle}{\left\|\mathscr{P}_h^{-1/2}(\nabla z\cdot\mathbf{n})\right\|_{-1/2,\Gamma}}\\&\ge\overline{C}\frac{1}{\|\lambda_{\tilde{h}}\|_{1/2,\Gamma}}\left|\left\langle\nabla z\cdot\mathbf{n},\lambda_{\tilde{h}}\right\rangle-\left\langle\nabla z\cdot\mathbf{n}-\mathscr{P}_h^{-1/2}(\nabla z\cdot\mathbf{n}),\lambda_{\tilde{h}}\right\rangle\right|\\&\ge\overline{C}\frac{1}{\|\lambda_{\tilde{h}}\|_{1/2,\Gamma}}\left\{\hat{C}\,\|\lambda_{\tilde{h}}\|_{1/2,\Gamma}^2-C\left(\frac{h_\Gamma}{\tilde{h}}\right)^{\delta}\|\lambda_{\tilde{h}}\|_{1/2,\Gamma}^2\right\}\\&\ge\left\{C_1-C_2\left(\frac{h_\Gamma}{\tilde{h}}\right)^{\delta}\right\}\|\lambda_{\tilde{h}}\|_{1/2,\Gamma},\end{aligned}$$

from which, taking $C_0=\left(\dfrac{C_1}{2C_2}\right)^{1/\delta}$, the proof is concluded. □

We notice now that, to have the inverse inequality for $Q_{\tilde{h}}^{\xi}$, one needs to see that

$$\|\lambda_{\tilde{h}}\|_{1/2,\Gamma}\le\tilde{h}^0\,\|\lambda_{\tilde{h}}\|_{1/2,\Gamma}\quad\forall\lambda_{\tilde{h}}\in Q_{\tilde{h}}^{\xi}$$

and then prove that

$$\|\lambda_{\tilde{h}}\|_{1,\Gamma}\le C\tilde{h}^{-1/2}\,\|\lambda_{\tilde{h}}\|_{1/2,\Gamma}\quad\forall\lambda_{\tilde{h}}\in Q_{\tilde{h}}^{\xi},$$

for which it suffices to demonstrate that $|\lambda_{\tilde{h}}|_{1,\Gamma}\le C\tilde{h}^{-1/2}\,\|\lambda_{\tilde{h}}\|_{1/2,\Gamma}\quad\forall\lambda_{\tilde{h}}\in Q_{\tilde{h}}^{\xi}$.

PROCEDURE 2. Let $\Phi_h(\Gamma):=\left\{\phi_h\in L^2(\Gamma):\phi_h|_e\in\mathbb{P}_0(e)\quad\forall e\in\Gamma_h\right\}$, define $h_\Gamma:=\max\left\{|e|:e\in\Gamma_h\right\}$, and assume that the number of edges e of Γ_h is even. Then we set

$$Q_{\tilde{h}}^{\xi}:=\left\{\lambda_h\in C(\Gamma):\quad\lambda_h|_e\in\mathbb{P}_1(e)\quad\forall e\in\Gamma_{2h}\right\},$$

where Γ_{2h} is the partition of Γ that arises by joining adjacent edges (certainly lying on the same line). Note that $\dim \Phi_h(\Gamma) = 2 \dim Q_h^{\xi}$. As was the case earlier, the goal here is to show the existence of $\beta > 0$ such that

$$\sup_{\substack{\phi_h \in \Phi_h(\Gamma) \\ \phi_h \neq 0}} \frac{\langle \phi_h, \lambda_h \rangle}{\|\phi_h\|_{-1/2,\Gamma}} \geq \beta \, \|\lambda_h\|_{1/2,\Gamma} \qquad \forall \lambda_h \in Q_h^{\xi}. \tag{4.59}$$

To this end, we assume henceforth (see details in [39]) that there exist $\hat{\Phi}_h(\Gamma) \subseteq \Phi_h(\Gamma)$ and constants $\beta_0, \beta_1 > 0$, such that $\dim \hat{\Phi}_h(\Gamma) = \dim Q_h^{\xi}$,

$$\sup_{\substack{\phi_h \in \hat{\Phi}_h(\Gamma) \\ \phi_h \neq 0}} \frac{\langle \phi_h, \lambda_h \rangle}{\|\phi_h\|_{0,\Gamma}} \geq \beta_0 \, \|\lambda_h\|_{0,\Gamma} \qquad \forall \lambda_h \in Q_h^{\xi}, \tag{4.60}$$

and

$$\sup_{\substack{\phi_h \in \hat{\Phi}_h(\Gamma) \\ \phi_h \neq 0}} \frac{\langle \phi_h, \lambda_h \rangle}{\|\phi_h\|_{-1,\Gamma}} \geq \beta_1 \, \|\lambda_h\|_{1,\Gamma} \qquad \forall \lambda_h \in Q_h^{\xi}, \tag{4.61}$$

so that (4.59) is proved next by "*interpolating*" (4.60) and (4.61). In fact, we have the following result (cf. F.J. Sayas 2012, private communication).

Lemma 4.12. *There exists $\beta > 0$, independent of h, such that* (4.59) *holds.*

Proof. Let $G_h : L^2(\Gamma) \to Q_h^{\xi}$ be the operator defined by $G_h(\lambda) := \lambda_h$ for each $\lambda \in L^2(\Gamma)$, where λ_h is the unique element in Q_h^{ξ} such that

$$\langle \phi_h, \lambda_h \rangle = \langle \phi_h, \lambda \rangle \qquad \forall \phi_h \in \hat{\Phi}_h(\Gamma).$$

Note that the inf-sup condition (4.60) and the fact that $\dim \hat{\Phi}_h(\Gamma) = \dim Q_h^{\xi}$ guarantee the existence and uniqueness of λ_h. Observe, in addition, that λ_h is a Petrov–Galerkin type approximation of λ. Now it is easy to see that (4.60) and (4.61) can be put together in the form

$$\sup_{\substack{\phi_h \in \hat{\Phi}_h(\Gamma) \\ \phi_h \neq 0}} \frac{\langle \phi_h, \lambda_h \rangle}{\|\phi_h\|_{-s,\Gamma}} \geq \beta_s \, \|\lambda_h\|_{s,\Gamma} \qquad \forall \lambda_h \in Q_h^{\xi}, \qquad \forall s \in \{0,1\}.$$

Applying the preceding inequality to $G_h(\lambda) \in Q_h^{\xi}$, we obtain

$$\begin{aligned} \|G_h(\lambda)\|_{s,\Gamma} &\leq \frac{1}{\beta_s} \sup_{\substack{\phi_h \in \hat{\Phi}_h(\Gamma) \\ \phi_h \neq 0}} \frac{\langle \phi_h, G_h(\lambda) \rangle}{\|\phi_h\|_{-s,\Gamma}} \\ = \frac{1}{\beta_s} \sup_{\substack{\phi_h \in \hat{\Phi}_h(\Gamma) \\ \phi_h \neq 0}} \frac{\langle \phi_h, \lambda \rangle}{\|\phi_h\|_{-s,\Gamma}} &\leq \frac{1}{\beta_s} \|\lambda\|_{s,\Gamma} \quad \forall \lambda \in H^s(\Gamma), \end{aligned}$$

that is,

$$\|G_h(\lambda)\|_{s,\Gamma} \leq \beta_s^{-1} \|\lambda\|_{s,\Gamma} \qquad \forall \lambda \in H^s(\Gamma), \qquad \forall s \in \{0,1\},$$

and then, thanks to the interpolation estimates for Sobolev spaces (cf. [49, Appendix B]), we obtain

$$\|G_h(\lambda)\|_{1/2,\Gamma} \leq (\beta_0 \beta_1)^{-1/2} \|\lambda\|_{1/2,\Gamma} \qquad \forall \lambda \in H^{1/2}(\Gamma). \tag{4.62}$$

We now let $\mathscr{P}_h^{-1/2} : H^{-1/2}(\Gamma) \to \hat{\Phi}_h(\Gamma)$ be the orthogonal projector. Then, given $\lambda \in H^{1/2}(\Gamma)$, we consider the functional $f_\lambda : H^{-1/2}(\Gamma) \to \mathbb{R}$ defined by

$$f_\lambda(\phi) := \langle \mathscr{P}_h^{-1/2}(\phi), \lambda \rangle \qquad \forall \phi \in H^{-1/2}(\Gamma)$$

and let $v_\lambda := J^{-1}(f_\lambda) \in H^{1/2}(\Gamma)$, where $J : H^{1/2}(\Gamma) \to H^{1/2}(\Gamma)'' \equiv H^{-1/2}(\Gamma)'$ is the usual isometry characterizing reflexive spaces, that is, $J(v)(F) = F(v) \quad \forall F \in H^{-1/2}(\Gamma) \equiv H^{1/2}(\Gamma)', \quad \forall v \in H^{1/2}(\Gamma)$. It follows that $\forall \phi_h \in \hat{\Phi}_h(\Gamma)$ there holds

$$\begin{aligned}
&\langle \phi_h, G_h(v_\lambda) \rangle = \langle \phi_h, v_\lambda \rangle = J(v_\lambda)(\phi_h) = f_\lambda(\phi_h) \\
&= \langle \mathscr{P}_h^{-1/2}(\phi_h), \lambda \rangle = \langle \phi_h, \lambda \rangle = \langle \phi_h, G_h(\lambda) \rangle,
\end{aligned}$$

and therefore $G_h(v_\lambda) = G_h(\lambda)$. Consequently, utilizing (4.62) and the fact that $\|\mathscr{P}_h^{-1/2}(\phi)\|_{-1/2,\Gamma}$ is certainly bounded by $\|\phi\|_{-1/2,\Gamma}$, we find that for each $\lambda \in H^{1/2}(\Gamma)$ there holds

$$\begin{aligned}
(\beta_0 \beta_1)^{1/2} \|G_h(\lambda)\|_{1/2,\Gamma} &= (\beta_0 \beta_1)^{1/2} \|G_h(v_\lambda)\|_{1/2,\Gamma} \\
\leq \|v_\lambda\|_{1/2,\Gamma} &= \sup_{\substack{\phi \in H^{-1/2}(\Gamma) \\ \phi \neq 0}} \frac{|\langle \phi, v_\lambda \rangle|}{\|\phi\|_{-1/2,\Gamma}} \\
&= \sup_{\substack{\phi \in H^{-1/2}(\Gamma) \\ \phi \neq 0}} \frac{|\langle \mathscr{P}_h^{-1/2}(\phi), \lambda \rangle|}{\|\phi\|_{-1/2,\Gamma}} \\
&\leq \sup_{\substack{\phi \in H^{-1/2}(\Gamma) \\ \phi \neq 0}} \frac{|\langle \mathscr{P}_h^{-1/2}(\phi), \lambda \rangle|}{\|\mathscr{P}_h^{-1/2}(\phi)\|_{-1/2,\Gamma}} \\
&\leq \sup_{\substack{\phi_h \in \hat{\Phi}_h(\Gamma) \\ \phi_h \neq 0}} \frac{\langle \phi_h \, \lambda \rangle}{\|\phi_h\|_{-1/2,\Gamma}}.
\end{aligned}$$

Finally, applying the preceding expression to $\lambda_h \in Q_h^{\xi}$, noting in this case that $G_h(\lambda_h) = \lambda_h$, and recalling that $\hat{\Phi}_h(\Gamma) \subseteq \Phi_h(\Gamma)$, we conclude that

$$(\beta_0 \beta_1)^{1/2} \|\lambda_h\|_{1/2,\Gamma} = (\beta_0 \beta_1)^{1/2} \|G_h(\lambda_h)\|_{1/2,\Gamma}$$

$$\leq \sup_{\substack{\phi_h \in \Phi_h(\Gamma) \\ \phi_h \neq 0}} \frac{\langle \phi_h, \lambda_h \rangle}{\|\phi_h\|_{-1/2,\Gamma}} \leq \sup_{\substack{\phi_h \in \Phi_h(\Gamma) \\ \phi_h \neq 0}} \frac{\langle \phi_h, \lambda_h \rangle}{\|\phi_h\|_{-1/2,\Gamma}} \qquad \forall \lambda_h \in Q_h^\xi,$$

which constitutes the required discrete inf-sup condition (4.59). □

We now look at the discrete kernel V_h of b, that is,

$$V_h := \left\{ \tau_h \in H_h : \quad b(\tau_h, (v_h, \lambda_h)) = 0 \quad \forall (v_h, \lambda_h) \in Q_h \right\},$$

which, according to (4.51) and (4.54), yields

$$V_h := \left\{ \tau_h \in H_h : \quad \operatorname{div} \tau_h \in \mathbb{P}_0(\Omega) \quad \text{and} \quad \langle \tau_h \cdot \mathbf{n}, \lambda_h \rangle = 0 \quad \forall \lambda_h \in Q_h^\xi \right\}.$$

Hence, the V_h-ellipticity of a follows straightforwardly from [39, Lemma 3.2] by making use of only the first property characterizing the elements of V_h.

Consequently, applying again the discrete Babuška–Brezzi theory (cf. Theorems 2.4 and 2.6) we conclude that (4.49) has a unique solution $(\sigma_h, (u_h, \xi_h)) \in H_h \times Q_h$, and there exists a constant $C > 0$, independent of h, such that

$$\|(\sigma, (u, \xi)) - (\sigma_h, (u_h, \xi_h))\|_{H \times Q} \leq C \inf_{(\tau_h, (v_h, \lambda_h)) \in H_h \times Q_h} \|(\sigma, (u, \xi)) - (\tau_h, (v_h, \lambda_h))\|_{H \times Q}.$$

The approximation properties of H_h and Q_h^u are somehow already established by (4.32) and (4.33) [see also (4.4) and (4.21)], whereas that of Q_h^ξ is given by (cf. [39, **(AP3)**])

$$\|\lambda - \mathscr{P}_{1/2,h}(\lambda)\|_{1/2,\Gamma} \leq C h^\delta \|\lambda\|_{1/2+\delta,\Gamma} \quad \forall \lambda \in H^{1/2+\delta}(\Gamma), \quad \forall \delta \in [0,1],$$

where $\mathscr{P}_{1/2,h} : H^{1/2}(\Gamma) \to Q_h^\xi$ is the orthogonal projector with respect to the inner product of $H^{1/2}(\Gamma)$.

4.5 Linear Elasticity Problem

In this section we analyze the Galerkin scheme for the two-dimensional version of the linear elasticity problem with Dirichlet boundary conditions studied in Sect. 2.4.3.1. To this end, we recall that, given a bounded domain $\Omega \subseteq \mathbb{R}^2$ with Lipschitz-continuous boundary Γ, and given $\mathbf{f} \in \mathbf{L}^2(\Omega)$, the corresponding mixed formulation reduces to [cf. (2.50)] the following problem: find $(\sigma, (\mathbf{u}, \rho)) \in H_0 \times Q$ such that

$$\begin{aligned} a(\sigma,\tau)+b(\tau,(\mathbf{u},\rho)) &= F(\tau) \quad \forall \tau \in H_0, \\ b(\sigma,(\mathbf{v},\eta)) &= G(\mathbf{v},\eta) \quad \forall (\mathbf{v},\eta) \in Q, \end{aligned} \tag{4.63}$$

where (cf. Sect. 2.4.3.1)

$$H_0 := \left\{ \tau \in \mathbb{H}_0(\mathbf{div};\Omega) : \int_\Omega \operatorname{tr}(\tau) = 0 \right\}, \quad Q := \mathbf{L}^2(\Omega) \times \mathbb{L}^2_{\text{skew}}(\Omega),$$

$a : H_0 \times H_0 \to \mathbb{R}$ and $b : H_0 \times Q \to \mathbb{R}$ are the bilinear forms defined by

$$a(\zeta,\tau) := \int_\Omega \mathscr{C}^{-1}\zeta : \tau = \frac{1}{2\mu}\int_\Omega \zeta : \tau - \frac{\lambda}{4\mu(\lambda+\mu)}\int_\Omega \operatorname{tr}(\zeta)\operatorname{tr}(\tau) \tag{4.64}$$

for all $(\zeta,\tau) \in H_0 \times H_0$, and

$$b(\tau,(\mathbf{v},\eta)) := \int_\Omega \mathbf{v}\cdot\mathbf{div}\,\tau + \int_\Omega \eta : \tau \tag{4.65}$$

for all $(\tau,(\mathbf{v},\eta)) \in H_0 \times Q$, and the functionals $F \in H_0'$ and $G \in Q'$ are given by

$$F(\tau) := 0 \quad \forall \tau \in H_0, \qquad G(\mathbf{v},\eta) := -\int_\Omega \mathbf{f}\cdot\mathbf{v} \quad \forall (\mathbf{v},\eta) \in Q. \tag{4.66}$$

We now let $\{H_h\}_{h>0}$, $\{Q_{1,h}\}_{h>0}$, and $\{Q_{2,h}\}_{h>0}$ be families of arbitrary finite element subspaces of H_0, $Q_1 := \mathbf{L}^2(\Omega)$, and $Q_2 := \mathbb{L}^2_{\text{skew}}(\Omega)$, respectively. Then, denoting $Q_h := Q_{1,h} \times Q_{2,h}$, we consider the associated Galerkin scheme: find $(\sigma_h,(\mathbf{u}_h,\rho_h)) \in H_h \times Q_h$ such that

$$\begin{aligned} a(\sigma_h,\tau_h)+b(\tau_h,(\mathbf{u}_h,\rho_h)) &= F(\tau_h) \quad \forall \tau_h \in H_h, \\ b(\sigma_h,(\mathbf{v}_h,\eta_h)) &= G(\mathbf{v}_h,\eta_h) \quad \forall (\mathbf{v}_h,\eta_h) \in Q_h. \end{aligned} \tag{4.67}$$

Our goal is to apply the theory from Sect. 2.5 to find specific subspaces H_h, $Q_{1,h}$ and $Q_{2,h}$ ensuring the unique solvability and stability of (4.67). More precisely, assuming in advance that a is going to be elliptic on the discrete kernel V_h of b (which actually will be shown later on), we concentrate in what follows on proving the discrete inf-sup condition for b, that is, the existence of $\beta > 0$, independent of h, such that

$$\sup_{\substack{\tau_h \in H_h \\ \tau_h \neq 0}} \frac{b(\tau_h,(\mathbf{v}_h,\eta_h))}{\|\tau_h\|_H} \geq \beta \|(\mathbf{v}_h,\eta_h)\|_Q \quad \forall (\mathbf{v}_h,\eta_h) \in Q_h. \tag{4.68}$$

To accomplish this, we know from Fortin's lemma (cf. Lemma 2.6) that it suffices to build a sequence of uniformly bounded operators $\{\Pi_h\}_{h>0} \subseteq \mathscr{L}(H,H_h)$ such that

$$b(\tau - \Pi_h(\tau),(\mathbf{v},\eta_h)) = 0 \quad \forall (\mathbf{v},\eta_h) \in Q_h. \tag{4.69}$$

We now let $b_1 : H_0 \times Q_1 \to \mathbb{R}$ and $b_2 : H_0 \times Q_2 \to \mathbb{R}$ be the bounded bilinear forms such that

$$b(\tau,(\mathbf{v},\eta)) = b_1(\tau,\mathbf{v}) + b_2(\tau,\eta) \quad \forall (\tau,(\mathbf{v},\eta)) \in H_0 \times Q,$$

that is,

$$b_1(\tau,\mathbf{v}) := \int_\Omega \mathbf{v}\cdot\mathbf{div}\,\tau \quad \text{and} \quad b_2(\tau,\eta) := \int_\Omega \eta : \tau.$$

It follows that (4.69) can be rewritten as

$$b_1(\tau - \Pi_h(\tau),\mathbf{v}_h) + b_2(\tau - \Pi_h(\tau),\eta_h) = 0 \qquad \forall (\mathbf{v}_h,\eta_h) \in Q_h. \tag{4.70}$$

In addition, if we assume for a moment that we already have a sequence of uniformly bounded operators $\{\Pi_{1,h}\}_{h>0} \subseteq \mathscr{L}(H,H_h)$ such that

$$b_1(\tau - \Pi_{1,h}(\tau),\mathbf{v}_h) = 0 \qquad \forall \mathbf{v}_h \in Q_{1,h}, \tag{4.71}$$

then we aim to find a second sequence of uniformly bounded operators $\{\Pi_{2,h}\}_{h>0} \subseteq \mathscr{L}(H,H_h)$ such that

(i) $b_1(\Pi_{2,h}(\tau),\mathbf{v}_h) = 0 \quad \forall \mathbf{v}_h \in Q_{1,h}$ and
(ii) $b_2(\tau - \Pi_{1,h}(\tau) - \Pi_{2,h}(\tau),\eta_h) = 0 \quad \forall \eta_h \in Q_{2,h}$,

so that defining $\Pi_h := \Pi_{1,h} + \Pi_{2,h}$ we satisfy the condition (4.70) [equivalently, (4.69)]. Indeed, it is easy to see from the analysis in Sect. 4.2 that, given a triangularization $\mathscr{T}_h$ of $\overline{\Omega}$ and an integer $k \geq 0$, and defining

$$H_h := \{\tau_h \in H_0 : \quad \tau_{h,i}|_K \in RT_k(K) \quad \forall K \in \mathscr{T}_h\} \tag{4.72}$$

and

$$Q_{1,h} := \{\mathbf{v}_h \in \mathbf{L}^2(\Omega) : \quad \mathbf{v}_h|_K \in [\mathbb{P}_k(K)]^2 \quad \forall K \in \mathscr{T}_h\}, \tag{4.73}$$

where $\tau_{h,i}$ stands for the ith row of τ_h, one can proceed by rows as in (4.29) to define a uniformly bounded explicit family $\{\Pi_{1,h}\}_{h>0}$ satisfying (4.71).

It remains therefore to build a family $\{\Pi_{2,h}\}_{h>0} \subseteq \mathscr{L}(H,H_h)$ verifying (i) and (ii). To this end, we now follow the approach introduced in [23] (see also [12] for further extensions). More precisely, we let X_h and Y_h be stable finite element subspaces for the usual primal formulation of the Stokes problem and, given $\tau \in H_0$, consider the associated Galerkin scheme: find $(\mathbf{z}_h, p_h) \in X_h \times Y_h$ such that

$$\begin{aligned} \int_\Omega \nabla\mathbf{z}_h : \nabla\omega_h + \int_\Omega p_h \operatorname{div}\omega_h &= 0 \quad \forall \omega_h \in X_h, \\ \int_\Omega q_h \operatorname{div}\mathbf{z}_h &= \int_\Omega (\tau - \Pi_{1,h}(\tau)) : S(q_h) \quad \forall q_h \in Y_h, \end{aligned} \tag{4.74}$$

where $S(q) := \begin{pmatrix} 0 & q \\ -q & 0 \end{pmatrix} \in Q_2 := \mathbb{L}^2_{\text{skew}}(\Omega) \quad \forall q \in L^2(\Omega)$. Note that the stability of (4.74) guarantees the existence of a constant $C > 0$, independent of h, such that

$$\|\mathbf{z}_h\|_{1,\Omega} + \|p_h\|_{0,\Omega} \le C\,\|\tau - \Pi_{1,h}(\tau)\|_{0,\Omega} \qquad \forall\,\tau \in H_0\,. \tag{4.75}$$

Now it is easy to see that

$$\int_\Omega q_h \,\mathrm{div}\,\mathbf{z}_h = \int_\Omega \mathbf{curl}\,\mathbf{z}_h : S(q_h),$$

where, denoting $\mathbf{z}_h := (z_{h,1}, z_{h,2})^{\mathrm{t}} \in X_h$,

$$\mathbf{curl}\,\mathbf{z}_h := \begin{pmatrix} -\dfrac{\partial z_{h,1}}{\partial x_2} & \dfrac{\partial z_{h,1}}{\partial x_1} \\ -\dfrac{\partial z_{h,2}}{\partial x_2} & \dfrac{\partial z_{h,2}}{\partial x_1} \end{pmatrix},$$

and hence the second equation in (4.74) can be rewritten as

$$\int_\Omega \Big(\tau - \Pi_{1,h}(\tau) - \mathbf{curl}\,\mathbf{z}_h\Big) : S(q_h) = 0 \qquad \forall\,q_h \in Y_h.$$

Thus, a comparison between this identity and the required condition (ii) for $\Pi_{2,h}$ suggests defining $\Pi_{2,h}(\tau) := \mathbf{curl}\,\mathbf{z}_h$ under the assumptions that $\mathbf{curl}\,(X_h) \subseteq H_h$ and $Q_{2,h} \subseteq S(Y_h)$. Next, since $\mathbf{div}\,\Pi_{2,h}(\tau) = \mathbf{div}\,\mathbf{curl}\,\mathbf{z}_h = 0$, it follows that

$$b_1\left(\Pi_{2,h}(\tau), \mathbf{v}_h\right) = \int_\Omega \mathbf{v}_h \cdot \mathbf{div}\,\Pi_{2,h}(\tau) = 0 \qquad \forall\,\mathbf{v}_h \in Q_{1,h},$$

which shows that (i) is also satisfied. In addition, thanks to the uniform boundedness of $\{\Pi_{1,h}\}_{h>0}$ [cf. (4.30)] and the stability result given by (4.75), we find that for each $\tau \in H_0$ there holds

$$\begin{aligned}
\|\Pi_{2,h}(\tau)\|_{\mathbf{div},\Omega} &= \|\mathbf{curl}\,\mathbf{z}_h\|_{0,\Omega} \le \|\mathbf{z}_h\|_{1,\Omega} \\
&\le C\,\|\tau - \Pi_{1,h}(\tau)\|_{0,\Omega} \le C\Big\{\|\tau\|_{0,\Omega} + \|\Pi_{1,h}(\tau)\|_{0,\Omega}\Big\} \\
&\le C\Big\{\|\tau\|_{0,\Omega} + C_1\,\|\tau\|_{\mathbf{div},\Omega}\Big\} \le C_2\,\|\tau\|_{\mathbf{div},\Omega}\,,
\end{aligned}$$

which proves that $\{\Pi_{2,h}\}_{h>0}$ is uniformly bounded as well.

As a consequence of the preceding analysis, we can say that, given a pair (X_h, Y_h) yielding a well-posed Galerkin scheme for the Stokes problem, the discrete inf-sup condition for b is insured by redefining

$$H_h := \Big\{\tau_h \in H_0 : \quad \tau_{h,i}|_K \in RT_k(K) \quad \forall\,K \in \mathcal{T}_h\Big\} + \mathbf{curl}(X_h)\,, \tag{4.76}$$

by keeping $Q_{1,h}$ as in (4.73), and by defining

$$Q_{2,h} := S(Y_h). \tag{4.77}$$

In particular, if we consider the "mini" finite element (cf. [41, Chap. II, Sect. 4.1]) given by

$$X_h := \left\{ \omega_h \in [C(\Omega)]^2 : \quad \omega_h|_K \in [\mathbb{P}_1(K) \oplus \langle b_K \rangle]^2 \quad \forall K \in \mathscr{T}_h \right\}$$

and

$$Y_h := \left\{ q_h \in C(\Omega) : \quad q_h|_K \in \mathbb{P}_1(K) \quad \forall K \in \mathscr{T}_h \right\},$$

where b_K is the bubble function on the triangle K, then (4.76) (with $k = 0$) and (4.77) become

$$H_h := \left\{ \tau_h \in H_0 : \quad \tau_{h,i}|_K \in RT_0(K) + \langle \mathrm{curl}\, b_K \rangle \quad \forall K \in \mathscr{T}_h \right\}$$

and

$$Q_{2,h} := \left\{ \eta_h := \begin{pmatrix} 0 & q_h \\ -q_h & 0 \end{pmatrix} : \quad q_h \in C(\Omega) \quad \text{and} \quad q_h|_K \in \mathbb{P}_1(K) \quad \forall K \in \mathscr{T}_h \right\},$$

which, together with $Q_{1,h}$ given by (4.73) (with $k = 0$), constitutes the well-known PEERS finite element subspace of order 0 for linear elasticity (cf. [2]).

Finally, it is easy to see from these definitions of H_h, $Q_{1,h}$, and $Q_{2,h}$ that the discrete kernel of b becomes

$$V_h := \left\{ \tau_h \in H_h : \quad \mathbf{div}\, \tau_h = 0 \text{ in } \Omega \quad \text{and} \quad \int_\Omega \eta_h : \tau_h = 0 \quad \forall \eta_h \in Q_{2,h} \right\}.$$

Hence, according to the inequalities given by (2.52) and Lemma 2.3, we conclude that a is V_h-elliptic, which completes the hypotheses required by the discrete Babuška–Brezzi theory (cf. Theorem 2.4) for the well-posedness of (4.67).

We end this monograph by mentioning that certainly many interesting topics have been left out of the discussion, including, to name just a few, *nonlinear boundary value problems, time-dependent problems, a posteriori error analysis, and further applications in continuum mechanics and electromagnetism* (see, e.g., the recent book [13] and the extensive list of references therein for a thorough discussion of them). In particular, it is worth mentioning that in the case of the linear elasticity problem, new stable mixed finite element methods in two and three dimensions with either strong symmetry or weakly imposed symmetry for the stresses have been derived over the last decade using the finite element exterior calculus, a quite abstract framework involving several sophisticated mathematical tools (e.g., [3–6]). In addition, concerning a posteriori error estimates for mixed finite element methods, we refer to the key contributions in [1, 17] and, within the context of the linear elasticity and Stokes problems, to [18, 19, 47]. Furthermore, and as complementary bibliographic material addressing some of the related contributions by the author, together with his main collaborators and former students, we may also refer to [10, 11, 21, 24, 25, 27–30, 32–34, 36–38], which deal mainly with *augmented mixed methods for linear and nonlinear problems in elasticity and fluid mechanics, twofold saddle point variational formulations, fluid–solid interaction problems, and the corresponding a posteriori error analyses*. We hope to write an extended version of the present book in the near future that will incorporate the contents of most of the aforementioned references.

References

1. Alonso, A.: Error estimators for a mixed method. Numer. Math. **74**(4), 385–395 (1996)
2. Arnold, D.N., Brezzi, F., Douglas, J.: PEERS: a new mixed finite element method for plane elasticity. Jpn. J. Appl. Math. **1**, 347–367 (1984)
3. Arnold, D.N., Falk, R.S., Winther, R.: Differential complexes and stability of finite element methods. II: the elasticity complex. In: Arnold, D.N., Bochev, P., Lehoucq, R., Nicolaides, R., Shashkov, M. (eds.) Compatible Spatial Discretizations. IMA Volumes in Mathematics and Its Applications, vol. 142, pp. 47–67. Springer, New York (2005)
4. Arnold, D.N., Falk, R.S., Winther, R.: Finite element exterior calculus, homological techniques, and applications. Acta Numer. **15**, 1–155 (2006)
5. Arnold, D.N., Falk, R.S., Winther, R.: Mixed finite element methods for linear elasticity with weakly imposed symmetry. Math. Comput. **76**(260), 1699–1723 (2007)
6. Arnold, D.N., Winther, R.: Mixed finite elements for elasticity. Numer. Math. **92**(3), 401–419 (2002)
7. Aubin, J.P.: Applied Functional Analysis. Wiley-Interscience, New York (1979)
8. Babuška, I., Aziz, A.K.: Survey lectures on the mathematical foundations of the finite element method. In: Aziz, A.K. (ed.) The Mathematical Foundations of the Finite Element Method with Applications to Partial Differential Equations. Academic, New York (1972)
9. Babuška, I., Gatica, G.N.: On the mixed finite element method with Lagrange multipliers. Numer. Meth. Partial Differ. Equat. **19**(2), 192–210 (2003)
10. Babuška, I., Gatica, G.N.: A residual-based a posteriori error estimator for the Stokes-Darcy coupled problem. SIAM J. Numer. Anal. **48**(2), 498–523 (2010)
11. Barrios, T.P., Gatica, G.N., González, M., Heuer, N.: A residual based a posteriori error estimator for an augmented mixed finite element method in linear elasticity. ESAIM Math. Model. Numer. Anal. **40**(5), 843–869 (2006)
12. Boffi, D., Brezzi, F., Fortin, M.: Reduced symmetry elements in linear elasticity. Comm. Pure Appl. Anal. **8**(1), 95–121 (2009)
13. Boffi, D., Brezzi, F., Fortin, M.: Mixed Finite Element Methods and Applications. Springer Series in Computational Mathematics, vol. 44. Springer, Berlin (2013)
14. Brenner, S.C., Scott, L.R.: The Mathematical Theory of Finite Element Methods, 3rd edn. Texts in Applied Mathematics, vol. 15. Springer, New York (2008)
15. Brezis, H.: Analyse Fonctionnelle: Théorie at Applications. Masson, Paris (1983)
16. Brezzi, F., Fortin, M.: Mixed and Hybrid Finite Element Methods. Springer, Berlin (1991)
17. Carstensen, C.: A posteriori error estimate for the mixed finite element method. Math. Comput. **66**(218), 465–476 (1997)
18. Carstensen, C., Causin, P., Sacco, R.: A posteriori dual-mixed adaptive finite element error control for Lamé and Stokes equations. Numer. Math. **101**(2), 309–332 (2005)

G.N. Gatica, *A Simple Introduction to the Mixed Finite Element Method: Theory and Applications*, SpringerBriefs in Mathematics, DOI 10.1007/978-3-319-03695-3,

19. Carstensen, C., Dolzmann, G.: A posteriori error estimates for mixed FEM in elasticity. Numer. Math. **81**(2), 187–209 (1998)
20. Ciarlet, P.: The Finite Element Method for Elliptic Problems. North-Holland, Amsterdam (1978)
21. Domínguez, C., Gatica, G.N., Meddahi, S., Oyarzúa, R.: A priori error analysis of a fully-mixed finite element method for a two-dimensional fluid-solid interaction problem. ESAIM Math. Model. Numer. Anal. **47**(2), 471–506 (2013)
22. Dupont, T., Scott, R.: Polynomial approximation of functions in Sobolev spaces. Math. Comput. **34**(150), 441–463 (1980)
23. Farhloul, M., Fortin, M.: Dual hybrid methods for the elasticity and the Stokes problems: a unified approach. Numer. Math. **76**(4), 419–440 (1997)
24. Figueroa, L.E., Gatica, G.N., Márquez, A.: Augmented mixed finite element methods for the stationary Stokes equations. SIAM J. Sci. Comput. **31**(2), 1082–1119 (2008)
25. Garralda-Guillem, A.I., Gatica, G.N., Márquez, A., Ruiz Galán, M.: A posteriori error analysis of twofold saddle point variational formulations for nonlinear boundary value problems. IMA J. Numer. Anal. doi:10.1093/imanum/drt006
26. Gatica, G.N.: Solvability and Galerkin approximations of a class of nonlinear operator equations. Zeitschrift für Analysis und ihre Anwendungen **21**(3), 761–781 (2002)
27. Gatica, G.N.: Analysis of a new augmented mixed finite element method for linear elasticity allowing $\mathbb{RT}_0$ - $\mathbb{P}_1$ - $\mathbb{P}_0$ approximations. ESAIM Math. Model. Numer. Anal. **40**(1), 1–28 (2006)
28. Gatica, G.N., Gatica, L.F.: On the a-priori and a-posteriori error analysis of a two-fold saddle point approach for nonlinear incompressible elasticity. Int. J. Numer. Meth. Eng. **68**(8), 861–892 (2006)
29. Gatica, G.N., Gatica, L.F., Márquez, A.: Analysis of a pseudostress-based mixed finite element method for the Brinkman model of porous media flow. Numer. Math. doi:10.1007/s00211-013-0577-x
30. Gatica, G.N., González, M., Meddahi, S.: A low-order mixed finite element method for a class of quasi-Newtonian Stokes flows. Part I: a-priori error analysis. Comput Meth. Appl. Mech. Eng. **193**(9–11), 881–892 (2004)
31. Gatica, G.N., Heuer, N., Meddahi, S.: On the numerical analysis of nonlinear two-fold saddle point problems. IMA J. Numer. Anal. **23**(2), 301–330 (2003)
32. Gatica, G.N., Hsiao, G.C., Meddahi, S.: A residual-based a posteriori error estimator for a two-dimensional fluid-solid interaction problem. Numer. Math. **114**(1), 63–106 (2009)
33. Gatica, G.N., Hsiao, G.C., Meddahi, S.: A coupled mixed finite element method for the interaction problem between an electromagnetic field and an elastic body. SIAM J. Numer. Anal. **48**(4), 1338–1368 (2010)
34. Gatica, G.N., Márquez, A., Meddahi, S.: Analysis of the coupling of primal and dual-mixed finite element methods for a two-dimensional fluid-solid interaction problem. SIAM J. Numer. Anal. **45**(5), 2072–2097 (2007)
35. Gatica, G.N., Márquez, A., Meddahi, S.: A new dual-mixed finite element method for the plane linear elasticity problem with pure traction boundary conditions. Comput. Meth. Appl. Mech. Eng. **197**(9–12), 1115–1130 (2008)
36. Gatica, G.N., Márquez, A., Rudolph, W.: A priori and a posteriori error analyses of augmented twofold saddle point formulations for nonlinear elasticity problems. Comput. Meth. Appl. Mech. Eng. **264**(1), 23–48 (2013)
37. Gatica, G.N., Márquez, A., Sánchez, M.A.: Analysis of a velocity-pressure-pseudostress formulation for the stationary Stokes equations. Comput. Meth. Appl. Mech. Eng. **199**(17–20), 1064–1079 (2010)
38. Gatica, G.N., Meddahi, S., Oyarzúa, R.: A conforming mixed finite element method for the coupling of fluid flow with porous media flow. IMA J. Numer. Anal. **29**(1), 86–108 (2009)
39. Gatica, G.N., Oyarzúa, R., Sayas, F.J.: Analysis of fully-mixed finite element methods for the Stokes-Darcy coupled problem. Math. Comput. **80**(276), 1911–1948 (2011)

40. Gatica, G.N., Sayas, F.J.: Characterizing the inf-sup condition on product spaces. Numer. Math. **109**(2), 209–231 (2008)
41. Girault, V., Raviart, P.-A.: Finite Element Methods for Navier-Stokes Equations: Theory and Algorithms. Springer, Berlin (1986)
42. Grisvard, P.: Singularities in Boundary Value Problems. Recherches en Mathématiques Appliquées, vol. 22. Masson, Paris; Springer, Berlin (1992)
43. Heuer, N.: On the equivalence of fractional-order Sobolev semi-norms. arXiv:1211.0340v1 [math.FA]
44. Hiptmair, R.: Finite elements in computational electromagnetism. Acta Numer. **11**, 237–339 (2002)
45. Howell, J.S., Walkington, N.J.: Inf-sup conditions for twofold saddle point problems. Numer. Math. **118**(4), 663–693 (2011)
46. Kufner, A., John, O., Fučík, S.: Function Spaces. Monographs and Textbooks on Mechanics of Solids and Fluids; Mechanics: Analysis, pp. xv+454. Noordhoff, Leyden; Academia, Prague (1977)
47. Lonsing, M., Verfürth, R.: A posteriori error estimators for mixed finite element methods in linear elasticity. Numer. Math. **97**(4), 757–778 (2004)
48. Márquez, A., Meddahi, S., Sayas, F.J.: Strong coupling of finite element methods for the Stokes-Darcy problem. arXiv:1203.4717 [math.NA]
49. McLean, W.: Strongly Elliptic Systems and Boundary Integral Equations. Cambridge University Press, Cambridge (2000)
50. Quarteroni, A., Valli, A.: Numerical Approximation of Partial Differential Equations. Springer, New York (1994)
51. Raviart, P.-A., Thomas, J.-M.: Introduction á L'Analyse Numérique des Equations aux Dérivées Partielles. Masson, Paris (1983)
52. Roberts, J.E., Thomas, J.-M.: Mixed and hybrid methods. In: Ciarlet, P.G., Lions, J.L. (eds.) Handbook of Numerical Analysis. Finite Element Methods (Part 1), vol. II. North-Holland, Amsterdam (1991)
53. Salsa, S.: Partial Differential Equations in Action. From Modelling to Theory. Springer, Milan (2008)
54. Schechter, M.: Principles of Functional Analysis. Academic, New York (1971)
55. Xu, J., Zikatanov, L.: Some observations on Babuška and Brezzi theories. Numer. Math. **94**(1), 195–202 (2003)

Index

G.N. Gatica, *A Simple Introduction to the Mixed Finite Element Method: Theory and Applications*, SpringerBriefs in Mathematics, DOI 10.1007/978-3-319-03695-3,